T0298458

Security Architecture – How & Why

RIVER PUBLISHERS SERIES IN DIGITAL SECURITY AND FORENSICS

Series Editors:

ANAND R. PRASAD
Deloitte Tohmatsu Cyber LLC in, Japan

R. CHANDRAMOULI
Stevens Institute of Technology, USA

ABDERRAHIM BENSLIMANE
University of Avignon France

The "River Publishers Series in Security and Digital Forensics" is a series of comprehensive academic and professional books which focus on the theory and applications of Cyber Security, including Data Security, Mobile and Network Security, Cryptography and Digital Forensics. Topics in Prevention and Threat Management are also included in the scope of the book series, as are general business Standards in this domain.

Books published in the series include research monographs, edited volumes, handbooks and textbooks. The books provide professionals, researchers, educators, and advanced students in the field with an invaluable insight into the latest research and developments.

Topics covered in the series include-

- Blockchain for secure transactions
- Cryptography
- Cyber Security
- Data and App Security
- Digital Forensics
- Hardware Security
- IoT Security
- Mobile Security
- Network Security
- Privacy
- Software Security
- Standardization
- Threat Management

For a list of other books in this series, visit www.riverpublishers.com

Security Architecture – How & Why

Tom Madsen

NNIT, Denmark

Routledge
Taylor & Francis Group

River Publishers

LONDON AND NEW YORK

Published 2022 by River Publishers
River Publishers
Alsbjergvej 10, 9260 Gistrup, Denmark
www.riverpublishers.com

Distributed exclusively by Routledge
4 Park Square, Milton Park, Abingdon, Oxon OX14 4RN
605 Third Avenue, New York, NY 10017, USA

Security Architecture – How & Why / by Tom Madsen.

Routledge is an imprint of the Taylor & Francis Group, an informa business

ISBN 978-87-7022-584-7 (print)
ISBN 978-10-0079-429-8 (online)
ISBN 978-1-003-33938-0 (ebook master)

While every effort is made to provide dependable information, the publisher, authors, and editors cannot be held responsible for any errors or omissions.

Contents

Preface

Security Architecture or Enterprise Information security architecture, as it was originally coined by Gartner back in 2006, has been applied to many things and different areas, making a concrete definition of Security architecture a difficult proposition. But having an architecture for the cyber security needs of an organization is important for many reasons, not least because having an architecture makes working with cyber security a much easier job since we can now build on a, hopefully, solid foundation. Developing security architecture is a daunting job, for almost anyone and in a company that has not had a cyber security program implemented before, the job becomes even harder. The benefits of having a concrete cyber security architecture in place cannot be overstated! The challenge here is that security architecture is not something that can stand alone, it absolutely must be aligned with the business in which is being implemented.

In this book, I hope to bring across to you the importance of and the benefits of, having security architecture in place. The book will be aligned with most of the sub frameworks in the general framework called SABSA or Sherwood Applied Business Security Architecture. SABSA is comprised of several individual frameworks and there are several certifications that you can take in SABSA, something I highly recommend if Security Architecture is something you would like to pursue as a career path. Aside from getting validation of your skills, SABSA as a framework is focusing on aligning the Security Architecture with the business and the business strategy. An important task in developing a security strategy! Each of the chapters in this book will be aligned with one or more of the components in SABSA, the components will be described as with the introduction to each of the chapters in this introduction.

I will be using examples throughout this book, to get my points across. These examples are based on technology from Microsoft and Cisco. This does not mean that I am recommending those vendors for your own architectures! These are just the vendors that I have chosen to specialize in, so that is the technology I am using with my own clients when consulting for clients in

Denmark. Denmark is a huge Microsoft and Cisco country; hence they are the vendors I am specializing in. Now, the chapter descriptions:

Chapter 1 – Why Security?

This chapter will help you make the argument to your business or organization, as to why having a security architecture in place is important as well as describe the benefits to the business, an important argument to make!

Chapter 2 – Why Architecture?

Here I will try to describe what architecture in this context is all about, especially how a formal architecture makes integration between systems and infrastructure-less complex and thus easier to secure in the architecture.

Chapter 3 – Security Architecture Model

This is the first chapter, where we dig into SABSA. Specifically, this chapter will detail the six sub frameworks that SABSA consists of and prepare you for the more detailed treatment in later chapters, of these subunits of SABSA.

Chapter 4 – Contextual Architecture

One of the core reasons for the effectiveness of SABSA is the business-oriented approach to security that is applied in this framework. This chapter will focus on how to align security architecture with the business needs and any regulations/compliance that needs to be considered as part of the architecture.

Chapter 5 – Conceptual Architecture

This is where we as security architects begin to add real value to an organization. Here we will begin the work of conceptualizing the solutions that will serve the business needs and make changes and adaptions later in the process much easier.

Chapter 6 – Logical Architecture

The logical part of the security architecture will follow naturally from the conceptual steps we did in the previous chapter. Now we will be looking into the functional and requirements and how these will fit into physical architecture steps in the next chapter.

Chapter 7 – Physical Architecture

Until now we have been looking at the more theoretical parts of SABSA, now the rubber meets the road with actual boxes of hardware and software. This is also the layer where we as architects look into the various data structures in use and the physical security requirements surrounding our architecture.

Chapter 8 – Component Architecture

This is the layer in the SABSA framework where the more specialized tools and components are located. This is also the chapter we I will be using examples from Microsoft and Cisco, to help you try and operationalize some of the more theoretical parts of this book.

Chapter 9 – Security Policy

Any security service, in security architecture, will need to be managed. How you go about this will contribute to the effectiveness of the service in the architecture as well as the overall effectiveness of the entire security architecture. Unfortunately, this is the step that is often overlooked or not taken seriously when implementing security architecture. In this chapter, I will give you some pointers and suggestions you can use in your own project.

Chapter 10 – Applied Security architecture with SABSA

This is where the rubber meets the road. In this chapter, I will walk you through some architecture examples using Cisco and Microsoft Azure. The examples will be based on a technology refresh of the entire networking infrastructure and migration of servers and applications to Microsoft Azure.

List of Figures

List of Tables

1

Why Security?

To understand 'Security Architecture' we must first make sure that you fully understand the meaning of security. It is a term that is used many times in many contexts and frequently with different meanings depending on the context.

In this chapter will provide you with a foundational understanding of security and how it fits in with Security Architecture.

1.1 Business Prevention

Cybersecurity has a bad reputation. If you, like me, have worked as an information system security professional in a business environment you know this only too well. When you walk into the room everyone groans. They say: 'Here come the security guys again! They are going to give us even more passwords to remember, more rules to enforce and they will create even more difficulties in our lives that will prevent us from getting on with real business. Why don't they just leave us alone' ?

I have even heard someone refer to the IT security organization as the 'business prevention' department! It is not an entirely unfair reputation. Are we being misjudged and slandered by our colleagues? Well, if we are honest with ourselves, as a profession we probably deserve some of it. But the profession has certainly got that reputation because we collectively behaved like that and did not understand the business environment that our recommendations and mandates had to fit inside.

How did we get this reputation? What did we do wrong? As I see it, we did not necessarily do anything directly wrong at the time this reputation developed. But as mentioned before, cybersecurity fits into a larger whole, inside the organization or business, that we are working within and understanding this environment requires that the security individual understand

1

this environment. Many cybersecurity professionals come from a technical background, without any business experience, something i am happy to see is slowly changing and business understanding is a core part of most of the educational efforts the schools and universities are offering on their cybersecurity programs.

1.2 Measuring and Prioritizing Business Risk

Security is used to protect assets with a value. If assets are in some way damaged or destroyed, then you will experience a business impact. The potential event by which you can suffer the damage or destruction is a threat, to prevent threats from crystallizing into a loss event that has a business impact, you use a protection or mitigation, measure to keep the threats away from your assets. If the assets are poorly protected, then you have a vulnerability to the threat. To improve the protection and reduce the vulnerability you introduce security controls, which can be either technical or procedural.

The process of identifying business assets, recognizing the threats, assessing the level of business impact that would be suffered if the threats were to materialize and analyzing the vulnerabilities, is known as a risk assessment and a risk assessment is not a one of exercise. Mature companies are conducting these kinds of assessments on a continuing basis and applying suitable controls to gain a balance between security, usability, cost and other business requirements as a part of their normal operations.

Risk assessment and risk mitigation jointly comprise what is often called operational risk management. Later chapters in this book examine operational risk management in much greater detail.

The main thing that you need to understand at this stage is that risk management is all about identifying and prioritizing the risks through the process and applying appropriate levels of control in line with those priorities.

Not all risks are worth the effort of implementing additional security and controls, either because the potential losses are not significant enough or because the costs of implementing the controls are higher than the value of the asset that is to be protected. What you get from a risk assessment is a set of business requirements for security, ranked in order of priority. These are most often expressed as a series of security and control objectives – abstract descriptions of business requirements for controls or mitigations. These in turn are used to drive the selection of risk mitigation approaches broad security and control strategies, logical security services, physical security mechanisms and finally the security products, tools and technology platforms with which you construct the Security Architecture.

Risk analysis comes in two forms, qualitative, where the risk assessments deliver a more subjective value of the various risks identified. A quantitative risk assessment delivers more concrete data values, that can be used by a company for prioritizing the efforts of protecting the various assets. These two forms are often used in conjunction with one another. The qualitative risk assessment provides the assessor with information on which risks might require a deeper analysis using quantitative methods, this is because using quantitative methods for all risks can be a massive investment in both time and money.

1.3 Security as a Business Enabler

The reputation that we, as information security professionals would like to have is quite different from the one that we have in many organizations. Although that is quickly changing, with the number of ransom ware attacks and data loss incidents we are seeing increases in frequency these years. Slowly but surely, this is changing to: Here come the security guys. They are going to help us to meet our business objectives and keep our data safe. Not the business prevention department, but the 'business enabling' department. But if we do our job properly and with due concern for the organization and the business environment that we are navigating within, we can make this happen. That is what our goal should be.

We must sell these information security ideas to our business colleagues and then make them come true. If we do not offer this sort of value to our business, then why are we there? There are several key technologies that are changing the way that business will be done in the future.

These include:

1. The Cloud
2. 5G mobile networking
3. Software-defined networking.
4. High bandwidth internet connections for the end-users
5. Wireless networking

The major change that we will see because of the deployment of these technologies is the continued migration of both the point of sale and the point of delivery right into the premises of the customer popularly known as the B2C (business-to-consumer) model.

People who want to buy something no longer need to make a physical visit to the vendor. They can use some of their communications technology to reach out from their home. They can browse through virtual shops, looking at virtual products on the virtual shelves. The products themselves may be picked automatically in the electronic warehouse, packed and sent to the customer with minimal human intervention. This same scenario applies to cases where business organizations. This is known as the B2B model. 'Supply chain management' and 'eProcurement' are among the most popular phrases used to describe the goals of business organizations when applying this model.

However, the number of threats, impacts and vulnerabilities that arise within all of these extremely complex systems is not to be trifled with. The major obstacle to the development of electronic business on such a huge scale is the low level of confidence that is inspired in the customer. Especially with the number of successful attacks we see increase these years.

Think of the business risks:

1. Disclosure of private, personal information,
2. Fraudulent buyers
3. Theft of credit card data
4. Errors and mistakes in such complex systems

So here is our opportunity to show how good we are. We have the whole world pleading for the security of information systems to enable them to do business and protect customer data. You have the technology to provide the solutions. What you must also demonstrate is that you have the associated skills to apply that technology to solve the problems facing the business.

You need much more than pure technology. You also need:

1. Good understanding of the business needs
2. Strategic architectures
3. Project management
4. Systems integration
5. Security management policies and practices
6. Enterprise-wide security culture and infrastructure

1.4 Empowering the Customers

We have looked at examples from the retail world of electronic commerce. In these cases, we see that electronic information systems are the means to empower the customer to gain greater benefits. These information systems, therefore, become important competitive factors for the suppliers, because the customers will use their power to select those suppliers who can meet the challenge of providing these benefits fastest and to the best price.

Information security is a critical component here, without it will be difficult for vendors to meet this customer service challenge. Customers will evaluate suppliers not only on the products themselves but also on how those products are marketed, sold and supported. Add to this the recent GDPR legislation from the EU, with this the customers are justifiably expecting us to protect their data as well. Losing customer data is a surefire way of ending up on the front pages of the newspapers and customers that see their data lost to the Internet are less likely to repeat business with us, making such losses life-threatening to a company!

Where online information systems are involved, that means that the quality, reliability, integrity and availability of those information services will be key factors in determining which suppliers succeed and which do not. Add to this confidentiality of the data we store on customers. To maintain that quality of service, one of the major tools you will need is an effective, risk-based information security program and a structured information systems security architecture, the very reason for this book!

1.5 Protecting Relationships

There is another security-related dimension to business relationships that we have not yet explored: the concept of trust. We shall return to this in detail later on in the book, but for now, let us take a quick glance at the subject.

When you do business with someone, at whatever level (personal or corporate), you are establishing some level of trust in the other party. You usually evaluate a number of signals that you receive, perhaps over some time, to determine how much you trust this person. How do they present themselves (Dress, Act, Personality?) Have they done business before? How did it go? How long has the company been established? Can you get a reference from someone else you know and trust (a trusted third party) – someone that already knows this person and can vouch for him or her? You know the drill!

Trust is an essential pre-requisite to doing business and trust is entirely a relationship thing. Trust is not created through IT systems but through some mutual knowledge between the parties. However, technical systems are used to protect the trust in the relationship that already exists. These technical services are no substitute for trust. They do not create trust. They merely protect the trust that already exists. However, indirect trust, through a third party (sometimes called transitive trust), is an important part of setting up digital business networks. It is obviously an advantage for both customers and suppliers to be empowered to do business with one another even though they have no previous knowledge of one another. This is where the third-party referee comes into the picture. The third-party needs to be trusted by both parties. This trusted third party is then able to play the role of 'introducer' by vouching for each of the two business parties to the other. This is usually achieved by the trusted third party issuing each entity with some certified credentials. This is called a digital certificate and is certified by a digital signature of a trusted third party. This is what we have been using for many years in the online space with digital certificates for the customer to be sure that the vendor they are interacting with is actually who they say they are.

It's like the situation where you go to a party at someone's house – someone who is an old friend of yours and with whom you have a long-standing trust relationship, built up through decades of experience and mutual interaction. At the party, another guest, someone who you have not met before, nor heard of, approaches you. It's quite different from meeting this person in a bar or on the street, where you might be very cautious and even suspicious of being approached by a stranger. The first thing you each ask one another is your name and how you know the host of the party. This establishes the credentials – 'Oh, I'm an old friend from college days' or 'I'm a work colleague'. It gives a new friendship a kick-start because you have established that you are both trusted by the host, who in this case acts as a trusted introducer for you both, giving both of you some confidence that it is alright to proceed with a friendship. You can begin to interact with a level of trust that would not be possible in the downtown bar. This is what a trusted third party is doing in the case of digital certificates. The third party is guaranteeing that the certificate can be trusted so we as consumers can trust the vendor exhibiting the certificate.

Many business deals are founded upon a personal introduction by a mutually trusted third party or by belonging to some business community that is in some way regulated by a trusted overseer.

So, when we build information systems, these technical systems can leverage the trust that already exists, whether directly or indirectly via the certificates, and they can protect those trusted business relationships in the course of doing business through this new information system-based medium.

1.6 To Summarize

Security is all about protecting business goals and assets. It means providing a set of controls that are matched to business needs and risk profiles, which in turn are derived from an assessment and analysis of business risks. The objective of risk assessment is to prioritize risks to focus on those that require mitigation.

Risk is a complex concept, and for any given course of action, there is a risk associated with doing that thing and risk associated with not doing it. Thus, one must take care not to mitigate a specific risk while unintentionally increasing the overall risk to the wider range of business goals and objectives. Something that is becoming increasingly more complex to do in an increasingly regulated world and the ever-increasing risk of a cyber-attack.

In its best possible light, security should be seen as enabling business by reducing risks to an acceptable level, thus allowing the business to make use of new technologies for greater commercial and information security advantage. Security can also be the means to add value to the core product by enabling information services that are essential to the enhancement of the product itself or to the operational support of the product out in the world, something I predict will become a business differentiator in the coming years.

Secure information services can empower the customers, enabling them to do business more easily and providing them with enhanced services that will have competitive value while ensuring that they trust in our efforts to protect them from harmful leaks of their data. Security in business information systems also protects and leverages the trust that exists between business partners, allowing them to establish relationships and to do business in new ways using new technologies. Technologies that might even open up new avenues of business for our organizations!

2

Why Architecture

This chapter explores what I mean by architecture. In particular, I will examine the differences between 'architecture' and, for lack of a better word, plumbing. Both of these areas provide great value to cybersecurity, but they are not the same thing. In the world of IT, people sometimes mix up which is which. In this chapter I will try to convey:

1. The concept of architecture is to integrate complex solutions to a diverse range of complex needs and to manage that complexity.
2. The layered approaches to architecture and the use of architectural reference models and frameworks.
3. The benefits of taking a strategic architectural approach as opposed to just applying solutions individually.

2.1 Origins of Architecture

Architecture originated in the building of towns and cities and everyone understands this meaning of the word, so it makes sense to me, to begin by examining the meaning of 'architecture' in this more traditional context. Architecture is a set of rules and conventions by which we create buildings that serve the purposes which they are intended for. An office building will look different from a residential home for instance. Our concept of architecture is one that supports ours needs to live, to work, to do business, to travel, to socialize and to pursue leisure activities. Architecture is founded upon understanding the needs that it must accommodate. These needs are expressed in terms of function, aesthetics, culture, government policies and regulations.

This all boils down to two major factors that determine what architecture we will create. These factors are:

1. The Purpose
2. Technological capabilities

2.2 Managing Complexity

One of the key functions of architecture as a product developed by the architect is to provide a framework and design within which complexity can be managed successfully. Small, isolated, individual projects do not need architecture, because their level of complexity is limited and the chief designer can manage the overall design. As the size and complexity of a project grow, however it becomes clear that more designers are needed, all working to create something that has the appearance of being designed by a single design authority. Also, if a project is not isolated in nature, but rather is intended to fit within a much larger and complex set of other projects, then architecture is needed to act as a road map which all these projects can be brought together into a more complete whole. The result must be as if they were all intended to be part of a single, large, project. This applies whether the various projects are designed and implemented simultaneously or if they are designed and implemented independently over an extended period. As complexity increases, then a framework is needed and will benefit the overall project program, within which each designer can work, contributing to the overall design.

2.3 Information Systems Architecture

The whole idea behind architecture in buildings has been adapted to areas other than the building of towns and cities. In more recent times the idea has been adopted in the context of designing and building computer systems and so the concept of information systems architecture was been born. Just like conventional architecture defines the rules and standards for the design of buildings, information systems architecture addresses the same issues for the design and construction of computers, communications networks and the distributed business systems that are implemented using the various technologies available to us. As with the conventional architecture of buildings and cities, information systems architecture must therefore consider the goals that are to be achieved with the systems we are designing. The technical skills of the people to construct and operate the systems and their individual sub-systems.

If we accept this foundation, then we are already well on our way to recognizing that information systems architecture is concerned with much more than just technical factors. It is concerned with what the enterprise wants to achieve and the business environment that these systems will have to fit within. Technical factors are often the main ones that influence the architecture, how often have you heard the argument that we know Microsoft or Oracle, hence these technologies must be used, and under these conditions, the architecture can fail to deliver business expectations. This book is mainly concerned with the security of the business information systems, although I will be touching on other areas around this core subject. Hence the focus is on an enterprise security architecture, to emphasize that it is the enterprise and its activities that are to be secured and that the security of the underlying infrastructure is only part of this overall goal.

2.4 Architectures

Security Architecture encompasses much different architecture and is touched by much different architecture within the modern enterprise. Below I will be touching on a few of these architectures, but keep in mind that there is a plethora of literature that is covering these areas in much more detail! This literature should be on your to-do list for a firmer foundation in enterprise architecture.

2.4.1 Business Architecture

The business architecture describes an enterprise-wide perspective on how the business itself is structured into an organizational model and a set of processes, functions governance and the like. This is the primary architecture for all the below architecture types. The other sub-architectures are all created to support this one single overriding framework of how the business works.

2.4.2 Information Architecture

Any business around the world is represented by information. Every business relationship, every business process, every business transaction, everything about the business, is represented by information. Information is the more abstract representation of something real and tangible, like the product that a business might produce. So, information is important to a business, because the information is the business! The information is represented and stored in information systems and applications. The information architecture describes how the information is created, organized, processed, stored, retrieved and communicated. Information architecture describes information types and the relationships and organization, information behavior, information management processes and physical locations and repositories for information. It identifies and describes the major categories of information that are needed to support the business.

2.4.3 Applications Architecture

The applications are the, in our day and age, a vast amount of computer applications that are assisting the modern organization by carrying out actions on business information on behalf of the business. The applications architecture describes how applications are designed, how they integrate with one another and how they are supported within the business infrastructure environment (hardware, software and communications networks). The applications must relate to the business processes that they support and the information resources that they create, maintain and the process by adhering to the information architecture above. Characteristics of modern applications architecture are likely to be: Cloud-based, Web API's, micro services and reusable, generic modules and hence quickly adaptable to new business needs; Built on a strategic ERP system or the like, offering distributed processing via the cloud. The main objective of applications architecture is to enable and automate business processes.

2.4.4 Infrastructure Architecture

The applications are running on both virtual and physical infrastructure. Infrastructure has been the area of focus for many years, regarding cyber security and it is an integral part of the Security Architecture as well. We will be returning to this in later chapters, but for now, it is defined as including: The computer platforms (hardware and operating systems); The computer networks (cables, lines, switches, routers, etc.); The layer of software that bridges between infrastructures that have different physical characteristics. This is commonly known as middleware and is becoming increasingly important with the steady increase in cloud deployments.

2.4.5 Risk Management Architecture

Risk Management Architecture is a concept that is crossing all the previous kinds of architecture. The model represented here is more a business model and not quite a systems model, although systems are a core ingredient in any kind of risk architecture. It is essential to see risk management as an activity that happens within all the layers and again, the risk is an excellent tool for communication with the business! This risk management architecture is close to the concept of Security Architecture, but it is not quite the same. I will return to the concept of risk management in several of the later chapters.

2.4.6 Governance Architecture

Surrounding all other components just mentioned is the all-important piece labeled Management and Governance Architecture. Governance is one of the things that I, personally, think will only become more important in the coming years. Good governance, whether it is IT governance or corporate governance, will become a differentiator for companies to show responsibility to both governments and customers, with all of the corporate scandals we have seen in recent years. The representation of this as an all-encompassing component of security architecture is important. It is through this framework that the senior management controls the business, manages risk and governs how the business uses information, applications and infrastructure. The management and governance architecture describes the decision-making processes and levels of authority that are assigned to decision-making entities (individuals or committees).

2.5 Enterprise Security Architecture

It is the common experience of many organizations that information security solutions are often designed, acquired and installed on a tactical basis without further thought to how this solution will fit within the already existing infrastructure and business environment. A requirement is identified, a use case is written and a solution is sought to meet the identified need. In this process, there is no opportunity to consider a more strategic dimension and the result is that the organization builds up a mix of various technical solutions on an ad hoc basis without integration between the components or the costs involved in operating a disparate set of security solutions.

Those enterprises that suffer these problems are often aware of these issues, but struggle to find an approach that will make things better. Good architecture never happens by accident and so the enterprise must find skills, methods and tools that help it to succeed with a more strategic architectural approach. One approach that avoids these problems is the development of an enterprise security architecture that is business-driven and describes a structured relationship between the technical and procedural solutions that supports the long-term needs of the business or organization. If the architecture is to be successful, then it must provide a framework within which decisions can be made on the selection of security solutions. These decision criteria absolutely must be derived from a thorough understanding of the business requirements, including the need for cost reduction, modularity, scalability, ease of re-use, operability, usability, inter-operability both internally and externally and integration with the enterprise ICT architecture and its legacy systems. Note the cost reduction part of the previous sentence, there is money to be saved here! Maybe not on the immediate implementation, but the long-term costs of the security solutions will be vastly decreased.

Furthermore, information system security is only a small part of information security, which in turn is just one part of a larger topic: business assurance. Business assurance consists of three major areas: information security; business continuity; physical and environmental security. Only through an integrated approach to these aspects of business assurance will it be possible for the enterprise to make the most cost-effective and beneficial decisions regarding the management of the various risks that any business or organization is facing daily. The enterprise security architecture and the security management process must therefore embrace all these areas. This brings us, finally, to SABSA.

2 SABSA® is a registered trademark of SABSA Limited. It stands for: Sherwood Applied Business Security Architecture

This book is a description of the SABSA® model and its applications. The model itself is described in much greater detail in Chapter 3. The primary characteristic of this model is that everything must be derived from an analysis of the business requirements for security, especially those in which security has an enabling function through which new business opportunities might be developed and exploited. The model is layered, with the top layer being the business requirements definition stage. At each of the lower layers, a new level of abstraction is developed, going through the definition of the conceptual architecture, logical architecture, physical architecture and finally at the lowest layer, the selection of technologies and products – or the shopping list.

The model itself is generic and can be the starting point for any kind of organization or business, by going through the process of analysis and decision-making implied by its structure, the output becomes specific to the enterprise and is at the end of the process customized to a unique business model. The output from applying the model becomes the enterprise security architecture and is central to the success of a strategic program of information security management within the organization.

2.6 Being a Successful Security Architect

Unless the security architecture can address a wide range of operational requirements and provide real business support and business enablement, rather than just focusing upon security, then it is likely that it will fail to deliver what the business expects and needs.

This type of failure is a common enough case throughout the information systems industry, not just within information systems security. In this book, I will put an enormous amount of emphasis on the need to avoid this mistake by keeping in mind the real needs of the business. It is not enough to compile a set of business requirements, document them and put them on the shelf, and then proceed to design a security architecture based on whatever technology the business is most familiar with.

Being a successful security architect means thinking in business terms, even when you get down to the real detail and nuts and bolts of the construction. You always need to have in mind the questions:

1. Why are we doing this?
2. What are we trying to achieve in business terms here?

It will also be difficult to battle against the numerous other people around you who do not understand strategic architecture and who think that it is all to do with technology. These people will constantly challenge you, attack you and ridicule you. You must be ready to deal with this.

You have to realize that being a successful architect is also about being a successful communicator/negotiator who can sell the ideas and the benefits to others in the enterprise that need to be educated about these issues. One of the most important factors for success is to have buy-in and sponsorship from senior management within the enterprise. Enterprise architecture cannot be achieved unless the most senior decision-makers are on your side.

Creating this environment of acceptance and support is probably one of the most difficult tasks that you will face in the early stages of your work.

2.7 Security Architecture Needs a Holistic Approach

Many people make the mistake of believing that building security into information systems is simply a matter of referring to a checklist of technical and procedural controls and applying the appropriate security measures said list. However, security has an important property that most people know about, but few pay any attention to. The security consists of many layers, or links in a chain, the chain is only as strong as the weakest link and an attacker will always go for this weakest link.

The checklist approach also fails because many people focus on checking that the links in the chain exist but do not necessarily that the links fit together to form a secure chain. The chain is a reasonably good analogy, but the problem is much worse. Imagine a checklist that has the following items: engine block, pistons, piston rings, piston rods, bearings, valves, camshaft, wheels, chassis, body, seats, steering wheel, gearbox, etc. Suppose that this list comprehensively itemizes every single component that would be needed to build a car or motorcycle. If you go through the checklist and make sure that you have all these components, does it mean that you have a car? Not even close!

Some of the key questions not addressed by the checklist approach to car construction are:

1. Can you be sure that all the parts have been designed to work together as one smoothly running system?
2. Do you have any assurance that the car has been properly assembled?
3. Has the engine been tuned?
4. Is the system running smoothly at this moment?

Checklists are never the entire answer. Security architecture, as with all other forms of architecture, needs a holistic approach, I know, another buzzword, but stay whit me:

1. Do you understand the requirements?
2. Do you have all the components?
3. Do these components work together?
4. Do they form an integrated system?
5. Does the system run smoothly?...

The analogy of the car as a complex machine that needs a holistic architectural design is much more powerful than the idea of a chain. Security architecture is more like the car, not the chain.

2.8 What Does Architecture Mean?

Architecture means taking a holistic, enterprise-wide view and creating principles, policies and standards by which the system (building, car, ship and business information system) will be designed and built and a firm foundation in the business needs.

The purpose of architecture is to ensure consistency of the design approach across a large complex system or across a complex array of smaller systems. Architectural approaches break up this complexity to present greater simplicity and thus make the various design activities easier to manage.

One of the ways to simplify complexity is to create architectural reference models that use layering of functionality to break down the complex whole into a series of less-complex conceptual layers, like the OSI model for networking.

The enterprise security architecture must be driven from a business perspective and business needs and must consider a wide range of requirements that are likely to conflict with one another. The successful architecture and architect, balance these tensions between the conflicting objectives.

Ad hoc approaches used instead of strategic thinking usually fail to satisfy the business needs or provide true business benefits. Enterprise security architecture needs a holistic system engineering approach that implies much more than simply satisfying all the points on a checklist.

The successful security architect is an experienced and intelligent person who is a good communicator and negotiator and can bring together many skills and wide-ranging knowledge from many parts of the team – someone who can grapple with the business requirements and use architectural skills to transform complexity into simplicity.

3

Security Architecture Model

The approach to developing an enterprise security architecture is based upon a six-layer model. The model is used as the basis for an architecture development process, or methodology. By following the development of the enterprise architecture in line with these layers of the model, the methodology becomes almost self-evident. Later chapters will provide guidance on how the individual steps and how they fit into the development of a security architecture.

In this chapter you will learn about:

1. The six-layered SABSA® Model and its relationship to the Zachman Framework (Zachman is a framework for enterprise architecture.)
2. The detailed description of each of the six layers of the SABSA® Model
3. The SABSA® Matrix shows the vertical analysis of each horizontal

3.1 The SABSA® Model

To establish a layered model of how security architecture is created and used, it is useful to return to the use of the word in its conventional sense: the construction of buildings. The SABSA® Model consists of six layers, see Table 3.1 for more details. It follows closely the work done by John A. Zachman, the creator of the Zachman framework, in developing a model for enterprise architecture, just adapted to a security view of the world and architecture in general. Each layer represents the view of a different player in the process of specifying, designing, constructing and using the building.

3.2 The Business View

When a new building is commissioned, the owner has a set of business requirements that must be met by the architecture. I for instance was part of

Table 3.1   The SABSA® model layers

The Business View	Contextual Architecture
The Architects view	Conceptual Architecture
The Designers View	Logical Architecture
The Builders View	Physical Architecture
The Consultants View	Component Architecture
The Managers View	Operational Architecture

the infrastructure requirements when a new building for the United Nations was built in Copenhagen Denmark back in the early 2010's. The team I was part of, was responsible for designing and coming up with the IT infrastructure requirements for the new building.

Each one of the business requirements implies an architecture that will be different from all the others and architecture that will fulfil expectations for the function of the building in business terms and needs. Having stated what sort of building is needed the owner must then decide some details about its use:

1. Why do you want this building?
2. How will it be used?
3. Who will use the building, including the types of people, their physical mobility, the numbers of them expected and so on?
4. Where should it be located, and what is its geographical relationship to other buildings and to the infrastructure (such as roads, railways, etc.)?
5. When will it be used? The times of day, week, year and the pattern of usage over time.

This type of analysis is essential before any type of design work is done. It is through this process that the requirements of the building are established and understanding the requirements is a prerequisite to designing a building that will meet those requirements.

When designing a secure business information system, the exact same principles apply. There are many possible architectural approaches that could be taken, but the one that will be the most suitable will be driven by a clear understanding of the business requirements for the system.

1. What type of information system is it and for what ends will it be used?
2. What is the system trying to achieve?
3. How will it be used?
4. Who will use it?

5. Where will it be used? Something that is becoming more important with the cloud becoming more and more important in architecture decisions!
6. When will it be used?

These are some of the questions that you must ask. From the analysis of the answers, you should be able to gain an understanding of the business requirements for the secure system. From these, you should be able to design systems architecture and a security architecture that meets those requirements.

In the SABSA® Model, this business view is called the *contextual security architecture*. It is a description of the business context in which your secure systems must be designed, built and operated. This context is extremely important for the further development and decisions on how the architecture is built and designed, so keep this in mind for the rest of the chapter.

Any attempt to define an architecture that takes a shortcut or avoids this essential step is unlikely to be successful. We continually see projects that fail to a degree that they end up in the newspapers because this essential step has been ignored. It is very common for systems architecture work to begin from a technical perspective, looking at technologies and solutions while ignoring the requirements.

It seems to be obvious common sense that one must first understand the requirements and still few people seem to know how to approach architecture development in the information systems arena. Unfortunately, many technologists and technicians believe that they already know the requirements, even though they have a poor relationship with those who might express these requirements, namely the business.

The results of taking a shortcut in the requirements-definition stages of an architecture development are abundantly clear. When one looks around at many large corporate enterprises and at their information systems infrastructure managers or applications teams, the relationship with the business community is often somewhat strained. For years the business people have been complaining that the information systems people are unable to deliver what the business needs and that ICT is a serious source of cost with little tangible benefit to show for it. The reason is simple: the business people are right. Vendor interests and technical innovations often drive the business systems development strategy, rather than it being driven by business needs. Those with responsibility for architecture and technical strategy often fail to understand the business requirements because they do not know how to do otherwise. This brings me back to some of my earlier points on the different languages that the business and ICT speak. Bridging this divide is of extreme importance currently, where the business has to move at a fast clip to keep up with competitors.

Ignorance of architectural principles is commonplace. This book aims to describe how to take a layered approach to security architecture development. Many of you will be tempted to flip through the pages to get to the end sections where some of the solutions can be found. You are in a hurry, I fully understand and whilst you know that this stepwise approach is correct, you simply do not have the time to linger on the details – you need to get to the meat. Well, be warned. There simply is no substitute for doing architecture work the proper way and that means attention to detail in this context. You may try to take shortcuts, but your efforts will most likely result in failure, which costs the business more money, delivers less benefit and destroys the confidence that business people may have in information and communications technology as the means to enable business development. If bad enough, it might be what is referred to as a 'resume generating event', a bad situation to be in!

In the model presented here, the contextual security architecture is concerned with:

1. *What*? The business, its assets to be protected and the business needs for information security
2. *Why*? The business risks in terms of assets, goals, success factors and the threats, impacts and vulnerabilities that put these at risk,
3. *How*? The business processes that require security
4. *Who*? The organizational aspects of business security
5. *Where*? The business geography and location-related aspects of business security
6. *When*? The business time dependencies and time-related aspects of business security in terms of both performance and sequence.

3.3 The Architect's View

The architect is someone creative with visions of how to achieve the end architecture. Good architects thrive on challenging business requirements and their real-world implications. They employ their skill, experience and expertise to create a vision of what the building will look like. They provide impressionistic drawings and high-level descriptions.

The visions are painted with broad strokes and lacking in detail. They prepare the way for more detailed work later, when other people with different types of expertise and skill will fill in the gaps with more fine brush strokes and details. The architect's view is the overall design that will meet the initial requirements. So, this layer of the architectural model is also referred to as *conceptual security architecture*. It defines principles and fundamental concepts which will guide the selection and organization of the logical and physical elements at the lower layers of the abstraction.

When describing an enterprise security architecture, this is the place to describe the security concepts and principles that you will use. These include:

1. What do you want to protect, expressed in the SABSA Model in terms of a SABSA Business Attributes Profile?

SABSA Business Attributes are explained in much greater detail in later chapters. They provide the primary tool by which business requirements can be captured in a formalized way.

1. *Why* the protection is important, in terms of the control objectives?

Control objectives are derived directly from an analysis of business risks and are a conceptualized way for business motivation for security.

1. How do you want to achieve the protection, in terms of technical and management security strategies?

These strategies set out the layered framework for integrating individual elements at the lower levels, ensuring that these fit together in a way that fulfils the overall goals of the business and the requirements. There are strategies for applications security, the network security strategy, the cryptographic infrastructure strategy, the role-based access control (RBAC) strategy, the cloud security strategy and so on. For every major area of the business requirements identified in the contextual security architecture, there will be a group of strategies designed to support it.

1. Who is involved in security management? a RACI (Responsible, Accountable, Consulted, Informed) chart is a good tool for this

The important trust concepts are concerned with the various policy authorities that govern trust within a domain, the policies that they set to govern behavior of entities in each of those domains, and the inter-domain trust relationships. So, yes, policies are an integrated part of any SABSA architecture, since policies are the go-to documentation for many requirements in an architecture.

1. Where do you want to achieve the protection in terms of the security domains?

The important concepts here are security domains (both logical and physical), domain boundaries and security associations.

1. When is the protection relevant? For instance, how long should data is encrypted?

The important concepts are lifetimes and expiration deadlines (of keys, certificates, passwords, sessions, etc.), and the use of trusted time for time stamping and time-sensitive business transactions.

3.4 The Designer's View

The designer takes over from the architect. The designer must interpret the architect's conceptual vision and turn it into a logical structure that can be engineered to create a real building.

In the world of computing and data communications, this design process is most often called systems engineering. It involves the identification and specification of the architectural elements of an overall system. This view models the business as a system, with system components that are themselves sub-systems that model and implement the detailed security strategies. It shows the major architectural security elements in terms of logical security services and describes the flow of control and the relationships between these elements.

In terms of architectural decomposition down through the layers, the logical security architecture must reflect and represent all of the major security strategies in the conceptual security architecture. At this logical level, everything from the higher layers is transformed into a series of logical abstractions. This will become clearer later when I walk through some examples in chapter 13.

The logical security architecture is concerned with:

1. What? It is the business information that needs to be secured.
2. Why? Specifying the security policy requirements for securing business information.
3. How? Specifying the logical security services and how they fit together.
4. Who? The users, security administrators, auditors, etc. and their inter-relationships.
5. Where? The security domains and inter-domain relationships.
6. When? The security processing cycle.

3.5 The Builder's View

The designer of the building hands over the process to the builder. It is the builder's job to choose and assemble the physical elements that will make the design come to life as a real construction. This view is also referred to as the physical security architecture.

In the world of business information systems, the designer produces a set of abstractions that describe the system to be built. These abstractions need to be turned into a physical security architecture model that describes the actual technology model and specifies the functional requirements of the various components. These logical security services are now expressed in terms of the physical security mechanisms and machines that will be used to deliver these services.

In total, the physical security architecture is concerned with:

1. What? The business data model and the security-related data structures
2. Why? The rules that drive logical decision-making within the system
3. How? The security mechanisms and the physical or virtual machines on which these mechanisms will be hosted.
4. Who? The people dependency in the form of the users, the applications that they use and the security user interface.
5. Where? The security technology infrastructure
6. When? The time dependency in the form of execution control structures (sequences, events, batches and the like).

Have you noticed that the questions repeat themselves in each of the layers we have discussed so far? The what, who, why, how, where and when are always good questions to ask in any new development efforts, just keep in mind that the business context must be part of the overall setting for the questions!

3.6 The Tradesman's View

When the builder plans the construction process, she or he needs to assemble a team of experts in each of the building trades that will be needed. It is the same for the construction of information systems. The builder needs to assemble a series of products from specialist vendors and a team with the skills to join these individual products together to form an implementation of the design.

Each of the integrators is the equivalent of a tradesman, working with specialist products and system components that are the equivalent of building materials and components. Some of these are hardware-related, some are software-related and some are service-oriented or cloud-oriented. The tradesmen work with a series of system components that are hardware items, software items or interface specifications and standards. Hence this layer of the architectural model is also called the component security architecture.

The component security architecture is concerned with:

1. *What*? Data field specifications, address specifications and other detailed data structure specifications.
2. *Why*? Security standards.
3. *How*? Products and tools (both hardware and software).
4. *Who*? User identities, privileges, functions, actions and access control lists (ACLs).
5. *Where*? Computer processes, databases, etc.
6. *When*? Security step timings and sequencing.

3.7 The Facilities Manager's View

When the building is finished, those who architected, designed and constructed it move out, but someone must run the building during its lifetime, a facilities manager. The job of the facilities manager is to deal with the operation and maintenance of the building. The framework for doing this is called the operational security architecture.

In the realm of business information systems, the operational architecture is concerned with classical systems operations work. Here the focus is only on the security-related parts of that work. The operational security architecture is concerned with the following:

1. *What*? Ensuring the operational continuity of the business systems and information processing and maintaining the security.
2. *Why*? To manage operational risks and hence to minimize operational failures and disruptions.
3. *How*? Performing specialized security-related operations
4. *Who*? Providing operational support for the security-related needs of all users and their applications
5. *Where*? Maintaining the integrity and security of all operational platforms and networks

Table 3.2 SABSA layer descriptions

Contextual Layer	Business policymaking, risk assessment process, business requirements collection and specification, organizational and cultural issues
Conceptual Layer	Major programs for training and awareness, business continuity management, audit and review, process development for registration, authorization, administration, incident response and development of standards and procedures
Logical Layer	Security policymaking, information classification, system classification, management of security services, security of service management and negotiation of inter-operable standards for security services, like some of the ISO standards
Physical Layer	Development and execution of security rules, practices and procedures, including, but not limited to: cryptographic key management, communication of security parameters between parties, synchronization between parties;
Component Layer	Products, technology, selection of standards and tools, project management, implementation, operation and administration of components

6. *When*? Scheduling and executing a timetable of security-related operations.

However, there is another dimension to the operational security architecture its relationship with the other five layers of the model. Thus, the operational security architecture needs to be interpreted in detail at each and every one of the other five layers. This is shown in Table 3.2, with some examples of the type of operational activity that is implied with regard to each of the layers.

3.8 The Inspector's View

There is another view of security in business information systems, the inspector's view (an IT auditor for instance), that is concerned with providing assurance that the architecture is complete, consistent, robust and fulfilling its purpose. In the realm of information systems security, this is the process of security auditing carried out by its auditors, systems quality assurance personnel or compliance auditors.

The SABSA Model does not see this as a separate architectural view. The SABSA approach to audit and assurance is that the architecture model as a whole supports these needs. The existence of such an architecture is one way in which the auditors will establish that security is being applied in a systematic and appropriate way according to the relevant policies and architecture details. The framework itself can provide a means by which to structure the audit process. In addition, security audit and review is addressed as one of the major strategic programs within the operational security architecture associated with the conceptual layer (see Table 3.1).

3.9 The Security Architecture Model

The SABSA Model for Security Architecture Development used in this book has six layers:

1. Contextual security architecture
2. Conceptual security architecture
3. Logical security architecture
4. Physical security architecture
5. Component security architecture
6. Operational security architecture

The operational layer should be visualized as a cross cutting concern across the other five layers, since there are operational aspects to all of these layers. Each of these six layers is further analyzed by asking six basic questions:

1. What?
2. Why?
3. How?
4. Who?
5. Where?
6. When?

Combining the layers with the analysis of the six questions, produces a 36-cell matrix, called the SABSA matrix. You can see an example of such a matrix in Table 3.3:

Table 3.3 SABSA Matrix

	Assets	Motivation	Process	People	Location	Time
Contextual	The Business	Business Risk Model	Business Process Model	Business Organization and Relationship	Business Geography	Business Time dependencies
Conceptual	Business Attributes Profile	Control Objectives	Security Strategies and Architectural Layering	Security Entity Model and Trust Framework	Security Domain Model	Security-Related Lifetimes and Deadlines
Logical	Business Information Model	Security Policies	Security Services	Entity Schema and Privilege Profiles	Security Domain Definitions and Associations	Security Processing Cycle
Physical	Business Data Model	Security Rules, Practices and Procedures	Security Mechanisms	Users, Applications and the User Interface	Platform and Network Infrastructure	Control Structure Execution
Component	Detailed Data Structures	Security Standards	Security Products and Tools	Identities, Functions, Actions and ACLs	Processes, Modes, Addresses and Protocols	Security Step Timing and Sequencing
Operational	Assurance of Operational Continuity	Operational Risk Management	Security Service Management and Support	Application and User Management Support	Security of Sites, networks and Platforms	Security Operations Schedule

I will be referring back to this table, in many places throughout this book, so I recommend that you bookmark this page for easy reference as you read through the book!

4

Contextual Security Architecture

The key to success in the SABSA® methodology is to be business-driven and business-focused. Like in all other security we absolutely must cater to the needs of the business! The business strategy, objectives, relationships and risks all tell you much about what sort of security architecture the organization needs. This analysis and description of the business itself are called contextual security architecture. This chapter will help you to focus on what it is you need to find out in order to construct your own contextual security architecture for your organization.

In this chapter I will elaborate on:

1. How information security can enable business activities that otherwise would be too risky
2. How digital business has developed and where it is going, and how good security is needed to protect the business activity
3. How all modern business activities, being so universally reliant on information and communications technology, require good information security
4. How safety-critical systems that use computers or electronics as part of their control logic need to be secured
5. The business goals, success factors and operational risks that drive the need for information security
6. Operational risk assessment is achieved through a process of risk modelling, threat assessment, business impact analysis and vulnerability analysis, followed by the prioritization of the risk identified
7. The types of information security services that need to be integrated into business processes
8. How organizational structures and business relationships affect the needs for information security.

4.1 Business Needs for Information Security

This section is relevant to the cell of the SABSA Matrix titled The Business on the Contextual row and in the Assets column (see Chapter 3, Table 3.3).

In any modern business environment, information and communications technology have become ubiquitous. It would be difficult to find an office-based job today that did not involve the employee using a computer system at some stage. Even in factories, workshops and construction sites, information and communications technology can be found. This tells us that:

1. Modern enterprises are totally dependent upon information and communications technology (ICT).
2. Modern enterprises are totally dependent upon the information.

To protect the information and its information processing capabilities, the enterprise needs to implement information security.

Information security has no intrinsic value of its own and is often seen as a useless expenditure at the business level. Its only possible value is that it protects something that has explicit value to the business. Therefore, you must begin the process of defining your business information security architecture by first identifying the things that you consider to be valuable that are affected by information security. These are the business assets that need to be protected by providing security for the business information and the information systems.

This chapter provides a checklist of items that you will want to investigate for your organization. It is not for this book to tell you how important these might be and you will need to investigate and prioritize them through meetings, interviews and document reviews in collaboration with the business.

From these initial business drivers, you then derive your Business Attributes, which in turn will drive the Business Risk Model, the Business Attributes Profile and the Control Objectives. The business drivers are the primary input to the process.

4.2 Security as a Business Enabler

This section is relevant to the cell of the SABSA® Matrix entitled The Business on the Contextual row and in the Assets column (see Chapter 3, Table 3.3). One of the key areas of the business need for information security is where there are new business activities enabled by information and communications technology or perhaps new ways of executing old business activities, again enabled by ICT.

4.2.1 On-Demand Entertainment

The whole sphere of multi-media technology now offers several ways in which music and movies can be delivered in digital format, directly online to a personal computer or digital TV. Once again, the issue facing such businesses is how to take advantage of this business opportunity without losing control over the materials.

The music industry faced something of a crisis over the electronic distribution of pirated digital copies made from CDs. The movie industry and the computer games industry both have similar issues.

4.2.2 Value-Added Information Services

We need information for all sorts of reasons to support our lifestyles. We need general news information, weather information, travel information, financial information, business information, sports news information, entertainment information and so on. Collecting, collating and presenting this information create a business opportunity. By adding value through personalized profiles, customized searches and the like, information providers develop a competitive edge for these services. However, for the business to be viable they need to sell their services and so they need to restrict access to authorized subscribers.

4.2.3 Remote Process Control

Electronics have been used in factories and manufacturing plants for some time. From a simple electronic control valve in a chemical plant to a fully automated robotic assembly line in a car manufacturing plant, there are countless applications of all shapes and sizes. These technologies are now being integrated with the traditional ICT systems in a big way, creating new challenges and security risks.

Potentially dangerous manufacturing processes can be controlled using standard technologies and protocols. A hacker can potentially intercept and hijack these remote communications and take over the plant. Something that has already happened to many companies. Now security becomes a major issue. What is worrying is that new adopters of these technologies are often blissfully unaware of these security issues and serious risks.

4.2.4 Supply Chain Management

Some businesses have complex supply chains and these need careful proactive management, just see the recent debacle that SolarWinds had with their supply chain!

Information and communications technology has impacted these kinds of complex businesses in two major ways. First, there are specialized computer applications for managing this complex database of information. The application shows current orders, history, seasonal patterns, etc., and in the more sophisticated versions, there is a facility for supply chain event management – providing visibility of the supply chain both upstream and downstream and reporting problems. Second, the communication of commercial information between suppliers and customers, such as orders, acknowledgments, invoices, remittance notices, etc., is often completely automated in electronic format.

4.2.5 Research and Information Gathering

The Internet has completely transformed research activities. Where libraries were once the only way of searching for information, online websites and sophisticated databases have replaced a vast proportion of these services that the traditional library used to provide. Scientific research and other areas of research and business intelligence gathering all now rely heavily upon electronic sources of information, often made available through the Internet.

These new ways of gathering information and searching for specific items are much faster than the old ways and provide a much wider scope for the searches, that is if the information is reliable, something that cannot be expected, unless we are using reliable sources for the information!

There is also an issue for the providers of authentic, specialized information about how to sell this service and charge subscription fees, so access control and authentication of authorized users is an issue for them.

4.3 Digital Business

This section is relevant to the cell of the SABSA Matrix titled The Business, on the Contextual row and in the Assets column (see Chapter 3, Table 3.3) Digital business is the term that emerged, largely promoted by The Gartner Group. However, digital business is slightly different in that

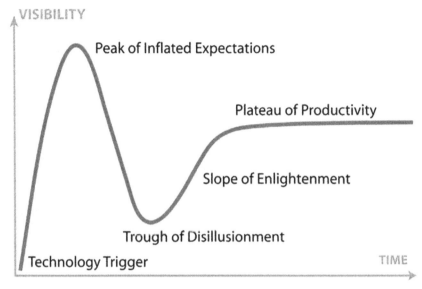

Figure 4.1   Event History

it takes a more realistic view that electronic business rarely exists in isolation and is in fact, just another kind of business, alongside the traditional bricks-and-mortar-based business.

Let's take a few moments to consider why the dot-com disaster happened at all because it has some bearing on what we are trying to do in building an enterprise security architecture. The history of the events is portrayed in Figure 4.1, which originally comes from Gartner but is taken from Wikipedia.

It shows a path of market enthusiasm from the original technology trigger of the Internet and the WWW, followed by a huge Peak of Inflated Expectations, down to the Trough of Disillusionment. What is expected after that is a gradual recovery of confidence based on more realistic expectations to a stable state in the Plateau of Profitability.

Many books have been written on the dotcom bust and subsequent time of self-reflection that came immediately after the slew of bankruptcies that came because of the excessive hype and investments. My own personal favorite of these books is one written by Andrew Smith, called Totally Wired, but there are others out there.

Despite the dotcom bust, quite a few positive business models and success stories appeared on the backend of the dotcom bust. The next section looks at a few of them.

4.3.1 Online Banking

Electronic banking is perhaps one of the most successful forms of electronic business. This success is probably due largely to the fact that the banking industry has always been some of the earliest adopters of information and communications technology and as an industry is therefore, the most experienced in its application to business development. This experience has paid off.

Banks have also had the longest commercial experience of applying information security. They deal almost exclusively in financial information, for which the security requirements are very clear, on top of that they are very heavily regulated, requiring additional compliance steps to be in place

Despite the level of experience that the banks have in offering and securing online services, the hackers are always coming up with new ways of attacking and the security architecture is sometimes not up to the challenge.

4.3.2 B2B

Supply chain management security has been discussed heavily in recent years, mainly due to the SolarWinds incident in December 2020. B2B is a sub-set of that discipline. B2B involves automated purchasing systems using software applications that are web or Internet-based. A business user with a common web browser can log into a procurement system to view vendor offerings and catalogues and place orders.

The process benefits can be:

1. Faster transmission of orders
2. Accurate order tracking
3. Efficient processing
4. Faster reconciliation
5. Lower transaction costs
6. Comprehensive reporting.

However, in companies that have implemented B2B, the major business benefit that they all report is quite different – than moving to B2B brings rogue buying under control overnight, forcing employees to stick to centrally negotiated procurement deals. They frequently report huge cost savings from this effect alone6, this means that the technology can be used to exert additional security and control over simple business processes where control has traditionally been difficult.

4.3.3 Online Government

Most governments in the western world are moving towards providing on-line services for citizens to interact with the many government departments and agencies. Typically, an individual must interact with 30 or 40 of such agencies over a lifetime and in most cases, he/she must submit all their personal information to each one. Online government services are therefore a tool for providing a more collaborative government, where once the government knows about you in one department, all other government departments also know the same information. This leads to greater efficiency in the government departments and less irritation for the citizen.

This integration of information across all departments is controversial. Some quite rightly see the dangers of moving towards a big brother state where the government can monitor your every move. Others point out the cost savings and the benefits to citizens in being able to get sensible replies and consistency in their interaction with the government. The security requirements for these eGovernment services are demanding. Authentication of the citizen is critical, especially as so many of them have the same names. In Denmark, where4 I live, we have a national ID number called a CPR number. This number is tied to a digital ID, that we use to interact with both the government and the various municipalities. Protection of privacy of personal information is also critical and the EU carries the weight of the GDPR laws behind it.

4.4 Continuity and Stability

This section is relevant to the cell of the SABSA Matrix entitled The Business, on the Contextual row and in the Assets column (see Chapter 3, Table 3.3). In addition to that, it also provides background information for developing the Business Model.

4.4.1 Revenue Generation

Any business that is dependent on computer systems to run its core business processes is dependent upon these systems to support the revenue stream. If these systems stop, then often so does the revenue. The continuity of service of such business systems is critical to the survival of the business. Questions that should be asked include:

1. How much money is lost if a system is unavailable?
2. How long will it take to recover the system?

3. What is the total revenue loss that would be accrued during recovery?
4. How long would it take before the business was in trouble with both cash flow?

Then look at all the threats that could bring down the service and decide what kind of architecture you need to protect the business from these risks.

4.4.2 Customer Service

A point that is often made in support of using corporate websites and other online information systems for customer interaction is that you can differentiate your business not only on the core product but on the levels of customer service that you wrap around it.

So, many enterprises have adopted technology to improve customer service. In doing so they have raised the expectations of the customers and once raised they are difficult to lower again. If you let the customer believe that you can offer a certain service level and you then fail to deliver that level, then the customer loyalty fades away fast. Remember, this is the Internet and a customer is only a few mice clicks away from changing their supplier.

If customer service is a key differentiator to your business, you will want to make sure that the security architecture supports the continuity and quality of service that you have promised your customers.

4.4.3 Reputation

Reputation is a difficult thing to quantify or describe in an objective manner. To some people, for an enterprise, it is reflected by the stock price, but this is too simplistic. So how can you measure reputation? You probably cannot find a hard numeric metric, so you must satisfy yourself with a qualitative approach.

It is certain that operational incidents will massively influence the reputation, but the question is, how much? A single incident or even a series of incidents of limited scope do not necessarily destroy the reputation of an organization but will affect it negatively! Reputation is a long-term thing. It takes years to build. Apart from certain catastrophic types of incidents, the reputation will survive a short-term battering. However, at some point in time short term begins to look like the long term. The transition point is difficult to predict, but once you get there, the reputation tends to flip very quickly to a bad one, a catastrophic failure where the collapse happens quickly and without warning.

Given that reputation failure exhibits this unpredictability, you must be a brave person to take risks with it. Most organizations that have a good solid reputation recognize that this is probably their single most important and most valuable asset, even though it does not appear on the asset register and its value cannot be seen on the balance sheet. So, protecting that reputation becomes one of the most critical of the business drivers.

In your organization, you need to ask around in senior management circles to get a feel for how reputation is viewed, how it is valued and how resistant it is thought to be in the face of damaging incidents. What do people think could cause material damage to the reputation of the organization? Then you will know how to treat it in prioritizing your business drivers.

4.4.4 Management Control

Keeping control of the business is the main job of management at all levels. In order to achieve that, the management must have information – 'data'. This is usually in the form of reports and analyses of how the business is performing. This information is usually generated by the business information systems.

The key security requirements for management information are integrity, is it correct? and timeliness. Confidentiality is also an issue to the extent that you probably do not want to share some of this information with outsiders just before a public announcement of an acquisition of a rival company. If yours is a company whose shares are quoted on the stock market then there are regulatory reasons to keep this management information confidential since it can lead to insider trading of your shares, which is illegal.

4.4.5 Operating Licenses

In some industries, you need a license to operate in business at all. These industries are mostly those where the health and safety of the public is an issue (like civil aviation, pharmaceuticals and health care) or where the security of the public's money is an issue (banking, insurance and investments). If you are in a regulated industry, then you will know about it. You will also know the regulatory and licensing conditions that are applied to your enterprise.

You need to analyze these business drivers very carefully to determine the extent to which they affect your information security requirements. Non-compliance can lead to censure, fines and at the end to the withdrawal of your operating license. You must decide how this will drive your enterprise security architecture.

4.4.6 Employee Confidence

There are multiple stakeholders in any modern enterprise and among the most important of these are the employees. This is especially true for companies in the knowledge economy, where the main company assets really are in the talents of the employees. So, in order to protect the strategic future of the enterprise, you must ensure that the employees have confidence in the organization and that they will remain loyal and happy to stay in your employment. I am surer that you have your own examples of employees walking out in protest of a #meeto culture in the company, as it happened at Blizzard, while I wrote this book.

Professional employees tend to be watchful and critical of the competence and ethics of their senior management. They look at the policies that are made, the personal examples that are set and most of all, they look to see whether the senior management seems to know what it is doing and whether it is really in control of the business. Overall lack of control, poor policymaking and inability to implement sensible policies eat away at employee confidence. Security management is an integral part of this picture and to maintain the confidence of the employees you need to make sure that you perform well in this area.

Set against that, the employees also want to feel they are trusted and that their requirements for ease of working have been considered. Thus, a security culture that treats employees as potential criminals and which makes their work difficult to execute will work against you. See the employees as part of the solution and not as the threat! That way you will garner trust with the employees and they will begin to suggest improvements, instead of working around the mitigation in place. There is a need for sensitivity and balance by ensuring that while free access to everything is not granted, employees find it easy to use the systems and they are empowered to do their work.

There is another aspect to employee confidence – that the employees feel protected against personal abuse. There are several themes in information system security that affect this:

1. Sexual harassment, which can happen through the e-mail system with suggestive remarks or pornographic images.
2. False accusations of computer misuse
3. Private personal information stored on corporate information systems must not be disclosed to unauthorized persons, either inside or outside the organization.

Reasonable personal use of corporate information systems and personal privacy are contentious issues, especially with respect to e-mail. Is the employee entitled to send private e-mails through the company system and if he or she does, is the company allowed to spy on these activities? The legal position differs from country to country and the moral position is also variable between cultures. These legal hot waters are more akin to boiling in many countries around the world. Make sure that the security architecture is designed to work with these hot topics!

Long ago, before the absolute pervasiveness of the telephone and before the age of the mobile telephone, the same issue surrounded the use of company telephones. Sensible companies allowed employees to make reasonable low-cost, short calls to arrange their personal life. Now there is the same debate about e-mail. It is clearly a reasonable thing for employees to communicate during the working day with family, friends and outside agencies to decide about their lives. The problem is where to draw the line since it is clearly unreasonable for an employee to spend a lot of time on this. It is also clearly unreasonable for an employee to spend a lot of time surfing the web for personal reasons or for an employee to use the company computers to run the accounting system for his or her private business. You must address these issues, decide what position you will take, set a Policy and publish clear guidelines so that people know exactly where they stand and implement the policy with fairness across the employee community.

4.4.7 Shareholder Confidence

Those who invest in your business need to be kept happy. Their happiness is tightly coupled with how well their investments are being managed, which aligns with how well the company is being managed. Shareholders are usually at arm's length with little direct visibility of company management. There are, however, several windows through which they look to see how things are going:

1. The annual report and accounts
2. The external auditor's annual report
3. The reports of the business and financial analysts

Shareholders are not concerned with the finer details, but they are concerned with the general ability of the management to manage the business, to deliver profits and to grow the assets. Corporate governance, the ability of the senior management to control the business, this is what gives shareholders confidence.

To maintain the confidence of shareholders your main aim must be to satisfy the external auditors and the analyst community and you can do this by making sure that your security management program satisfies the views of these groups. You should pay attention to audit points and take decisive action to satisfy the auditors and you absolutely must listen to what the analysts say and act accordingly.

4.4.8 Other Stakeholders

Not every organization is owned by shareholders. Some exceptions are government organizations and Non-Governmental Organizations (NGO's,) owned by the government and charitable organizations. Each model of ownership has its own specific set of stakeholders and you must decide what the ownership framework looks like in your organization. Whatever it is, it will have an impact on the needs of your enterprise security architecture.

For example, in government, a change of leadership or a change of party following an election will quickly lead to a change of policy and a major reorganization, with some departments being amalgamated, while others are being split, new departments being created and so on. The driver for these events is purely one of political presentation to the electorate, yet those responsible for organizing and managing information security must be able to respond flexibly and swiftly to the will of their political masters. If the security architecture is monolithic, then this will be very difficult, so a more finely granulated domain structure lends itself to the conceptual security architecture level. These business drivers must be understood.

4.5 Safety-Critical Dependencies

This section is relevant to the cell of the SABSA® Matrix entitled The Business, on the Contextual row and the Assets column (see Chapter 3, Figure 3.3).

Safety-critical systems are those whose failure may cause injury or death to human beings. Many such systems involve computer-controlled or electronic-controlled electromechanical sub-systems (SCADA, ICS, OT) and it is these that are of interest in this section.

4.5.1 Remote Communications to Safety-Critical Systems

One very important aspect of safety-critical systems has already been addressed earlier in this chapter about the need for securing remote control systems for factories, plants, robotic machinery and so on. Any application of process control that involves the use of remote data communications is at risk from an opponent intercepting and changing the communications and hijacking the control. This also applies to the remote management of computer systems and network devices and hence any business application supported on a remotely managed computer, cloud environment or data communications network is vulnerable to attack.

The means of securing remote communications to prevent these types of attacks lies in the use of cryptography to ensure complete authenticity and security of the instructions issued to the control target. Some protocols have standard optional features to implement these security mechanisms, but the ability to do so then depends upon vendor support for those mechanisms being built into the devices. Unfortunately, not all vendors are able to justify the extra cost of such support and in a competitive marketplace where the price is an issue, such extras are often not regarded as good business. This is changing, albeit slowly, with the realization that these kinds of systems are critical to the running of our societies!

There is an increasing tendency to introduce information and communications technology into complex engineering systems – like civil and military aircraft. Here safety-critical issues are paramount. Aircraft sub-systems are often engineered with triple redundancy to reduce the risk of failure. However, if remote digital data communications between ground stations and airborne flight vehicles are used, then security issues rear their heads yet again. The designers of such systems need to consider the authentication and security of this communication to ensure that a malicious opponent cannot hijack the aircraft by remotely hijacking the data communications.

4.5.2 Systems Assurance

Another important aspect of security regarding safety-critical systems is that of systems assurance. Here we are concerned with reaching high levels of assurance that the system has been implemented correctly and will function as expected. The areas of application include:

1. Power stations
2. Chemical plants, oil refineries and other manufacturing plants
3. Aircraft systems and air traffic control systems
4. Weapons and defense systems.

In most systems, it is enough to be sure that the system will perform all the functions laid down in the functional specification and that it is easy to test that it does. You simply write out the set of tests that covers all the functions, all the expected inputs and all the expected outputs and run through the tests to verify the validity of the system. In safety-critical systems, this is not sufficient, because the potential problems are not limited to the inputs and outputs but with unexpected situations. It is much more difficult to establish that the system does not have any unwanted reactions in response to unknown input data safety-critical security requirements are demanding and it remains an area in which there is much active research. Information and communications technology are being used more and more for these types of applications and there is a need to understand how to ensure that these systems do not endanger human life.

4.6 Business Goals, Success Factors and Operational Risks

This section is relevant to the cell of the SABSA® Matrix entitled Business Risk Model on the Contextual row and in the Motivation column (see Chapter 3, Table 3.3). The section looks at some of the key areas where the enterprise faces risk and is motivated to develop an information security response.

4.6.1 Brand Protection

A brand is something that takes a huge investment of time and money to develop. A brand is either a business name or product name, sometimes along with a pictorial image, which carries a message of reliability, quality and trustworthiness. It is very closely related to the reputation of a company or product. The brand is a major investment and therefore must be considered as a major asset, to be protected and further developed. Information security management plays an important part in this by supporting the overall strategic, tactical and operational development of the enterprise. Information security failures will lead to brand damage.

4.6.2 Fraud Prevention

Fraud is unfortunately widespread. It happens in every industry, in every size of an organization, at every level of the management hierarchy and it has existed for as long as there have been business ventures. So, a fraud carried out using a business information system is just another aspect of an age-old problem, the manipulating of business information to hide dishonesty and theft.

Computer-related fraud is perpetrated by abusing business computer systems that support business transactions (banking, procurement, etc.) or those that represent business positions (stock control, financial accounting, etc.). The effects of these frauds can be large or small, from large-scale bank transfer frauds, down to fiddling a travel expenses claim. They also vary from the single fraud to the systematic fraudulent collection of small, unnoticed.

Computers do not commit fraud. They are merely one of the many tools that can be used by humans to commit fraud and in that respect, computer-related fraud is just the same as any other type of fraud. Fraud usually happens in situations where there is an opportunity (access, skill and time) combined with motivation. Fraud prevention is relevant to every single business in the world and information security is needed to prevent the abuse of business information systems to commit fraud.

4.6.3 Loss Prevention

Potential business losses arise from a plethora of different operational risk areas. Information security management is one of the key competence areas in any organization that will help to manage and mitigate a wide range of these risks. Unfortunately, comprehensive risk management is lacking in many companies and organizations, even though such management can massively help with the reduction of risk and focus investments on areas where it will benefit the most. There is a comprehensive taxonomy of operational risks in Table 9.2 in a subsequent section of this chapter, together with their mapping to the information security domain.

4.6.4 Business Continuity

Perhaps the most experienced and most feared risk regarding using computers in business is the system failure that results in interruption of business operations. It can lead to delays or in some cases complete failure to meet the service level expectations of customers, suppliers, employees, shareholders, regulators, etc. If key business information services are disrupted, then so are the business processes that depend upon them.

Service interruptions can be caused by accidental system failures, by willful neglect or a lack of good administrative practices, or by malicious interference and sabotage. The entire security architecture is focused on upholding the set of business requirements that can be collectively classified under 'business continuity.' These requirements are common to every business at ever increasing levels of criticality, as IT systems are becoming ever more ubiquitous.

4.6.5 Legal Obligations

Organizations have many legal obligations and failure to meet these obligations represents a major area of operational risk. The situation is much more complex for companies who are international in their business operations since each operating country has different laws, some of which are in conflict across national borders. Many of the laws and regulations have a direct or indirect impact on the management of information security. It is not possible to list all the relevant laws and regulations for all the countries and all the industries, but here is a list of some of the important areas that you need to examine in your domains of operation:

1. Criminal law
2. Civil law
3. Regulatory compliance relevant to the industry
4. Contractual obligations
5. Management and mitigation of legal liabilities

In order to make sure that your enterprise security architecture considers all the legal and regulatory drivers, you will need to build a Business Attribute Profile that specifies the attributes and metrics to describe your business needs. In doing this you will need to confer not only with senior managers but also with the legal representatives who can give detailed advice. In certain regulated industries (like banking) there may also be a compliance officer or a compliance department that can give detailed advice on regulatory matters.

4.7 Operational Risk Assessment

This section is relevant to the cell of the SABSA® Matrix entitled Business Risk Model on the Contextual row and in the Motivation column (see Chapter 3, Table 3.3). The section looks at the basic modelling of risk and how such models can be used to carry out a risk assessment. This will help you to develop your Business Risk Model.

Risk Modelling

Risk is a complex concept, familiar to everyone in every aspect of daily life, yet surprisingly difficult to describe without a theoretical analysis. The most accepted model for risk involves some basic concepts:

1. Assets – things that are of value that you want to protect
2. Threats – potential damaging events that put your assets in danger

3. Impacts – the potential outcome of a threat materializing and causing damage to your assets
4. Vulnerabilities – weaknesses in your operational business procedures or systems that will allow a threat to materialize

The likelihood of a risk event occurring is some complex combination of:

1. Level of threat (the likelihood of the threat)
2. Level of vulnerability or weakness (the likelihood that a threat event will succeed in exploiting your business assets).

4.7.1 Risk/Threat Assessment

In order to manage risk, you first need to identify the sources of risk (threats) and assess their significance (the likelihood of the risk event and the impact on your business assets should it materialize). Risk assessment is a very important part of the SABSA® process (see also Chapter 3, Table 3.3). You will therefore need to adopt a risk assessment methodology to develop this business risk model.

Assessing the level of threat is notoriously difficult. Threats exist outside your span of control, the world is simply a dangerous place and all that you can do is to recognize the threats and their sources (threat agents). Without access to reliable, consistent, complete data on previous loss events, the statistical analysis provides little useful guidance on the probability of a threat materializing. Also, as has been already mentioned, observation of past events is not necessarily a good guide to how the future will be.

Assessing the vulnerabilities and the associated impact is much easier since both these things are within your span of control. Thus, risk assessment methodologies in commercial organizations tend to focus on assessing these aspects, usually qualitatively (low, medium, high). Analyzing threats in commercial organizations is limited simply to identifying the threats without quantification. Risk and threat assessment methodologies can fill several books by themselves and SABSA has its own model, but here I will point you to a few sources of risk and threat assessment methodologies and frameworks, before getting to the SABSA model. You absolutely must know about both risk and threat assessment, in order to be successful with SABSA!

1. ISO 31000 – Risk Management
2. ISO 27005 – Risk Management
3. COBIT 2019 – IT Risk Governance
4. ISACA Risk IT Framework.

Table 4.1   Threat domains and threat agents

Threat Domain	Description	Threat Agents
People	Malicious violation of internal policies	Current employees
	Negligent violation of internal policies	Past employees
	Human errors	People under consideration foremployment
Processes	Deficiency in an existing procedure	Employees
	Absence of a suitable procedure	Customers
	Failure to follow a defined procedure	Suppliers
		Service providers
		Agents
		Partners
		Members of the public
Systems	Unintentional losses caused by:	Technical failure
	Unforeseen breakdown of technical systems	Technical failure through inadequate
	Insufficient resilience in technical systems	design or poor implementation
External	Natural disasters	Natural events
	Man-made disasters (unintentional)	Accidents
	Malicious actions of third parties	Malicious third parties
	Negligent actions of third parties	Negligent third parties
	Legitimate actions of third parties	Legitimate third parties whose business
		interests' conflict with ours

4.7.2 Threat Domains

The definition of operational risk implicitly suggests that there are four principal domains of risk to be considered:

1. People
2. Processes
3. Systems
4. External events.

The SABSA® approach to risk assessment uses these domains as an initial classification for threats and threat agents that are relevant to operational risk assessment. Table 4.1 provides more detail.

4.7.3 Threat Categories

As a secondary classification of threats, the SABSA® approach uses a series of threat categories. The selection of these categories is entirely arbitrary and is based upon practical experience only. The categories chosen could be changed without violating any theoretical principles. Table 4.2 shows the taxonomy of threats classified by domains and categories. Each category maps to one or more of the threat domains. There is no specific logical mapping of a category to the domains. The category and domain mappings are indicated in columns 2– 5 in Table 4.2. The list is not necessarily comprehensive and new categories could be added as needed if that becomes necessary or desirable.

Table 4.2 Taxonomy of threats

Threat Category	Description	Security Mapping
Facilities and Environment	Loss or damage to operational capabilities caused by problems with premises, facilities, services or equipment.	Inadequate business continuity management and ICT disaster recovery
Health andSafety	Threats to the personal health and safety of staff, contractors, suppliers, agents, customers and members of the public	Inadequate protection for process control systems for automated factory processes, machine tools, power generation, etc., including automated secure vaults and other specialized machinery
Information Security	Unauthorized disclosure or modification to information or loss of availability of information or inappropriate use of information	Weak logical security, weak physical security and weak operating procedures
Control Frameworks	Inadequate design or performance of the existing risk management infrastructure	Inadequate measurement and reporting efficiency of systems and processes
Legal andRegulatoryCompliance	Failure to comply with the laws of the countries in which business operations are carried out or failure to comply with any regulatory, reporting and taxation standards or failure to comply with contracts, or failure of contracts to protect business interests	Inadequate architecture to enable upgrade, extension, change, enhancement of ICT systems to deal with changes in regulations and inadequate early warning research
Business Strategy	A strategic business plan fails to meet its expected targets	Poor market research for ICT services and weak forecasting techniques
Corporate Governance	Failure of directors to fulfil their personal statutory obligations in managing the company and protecting the interests of shareholders	Lack of due and timely diligence, inadequate reporting to senior management and lack of leadership in solving operational problems
Public Relations	The negative effects of public opinion, customer opinion, market reputation and the damage caused to the brand by failure to manage public relations	Inadequate crisis management, including public relations management

Continued

Table 4.2   Continued

Processing and Transactions	Problems with service or product delivery caused by failure of internal controls, information systems or through weaknesses in operating procedures	Inadequate exception reporting and transaction recovery in ICT systems
Product Liability	A product or service sold and delivered fails to meet the required standards for suitability for the client needs	Inadequate attention to fulfilling the SLA for ICT services
Behavioral	Problems with service or product delivery caused by lack of employee integrity, or by errors and mistakes	Inadequate ICT operations procedures and operations management
Technology	Failure to plan, manage and monitor the performance of technology-related projects, products, services, processes, staff and delivery channels	Inadequate ICT systems architecture to ensure flexibility in response to changing business requirements
Project Management	Failure to plan and manage the resources required for achieving tactical project goals, or leading to failure to complete the project	Need to ensure that ICT projects can be successfully be integrated with the business operations
Criminal Acts	Loss or damage caused by fraud, theft, willful neglect, gross negligence, vandalism, sabotage, extortion, etc.	Inadequate protection of private and confidential information gained from unauthorized access to ICT systems
Human Resources	Failure to recruit, develop or retain employees with the appropriate skills and knowledge or to manage employee relations	Poor transparency of processes and lack of cross training for people to deputize and step in for other key workers
Supply Chain	Failure to evaluate adequately the capabilities of suppliers leading to breakdowns in the supply process or sub-standard delivery of supplied goods and services; also failure to understand and manage the supply chain issues	Inadequate provision of strong contracts for ICT service suppliers addressing all aspects of security management and operations management, escalation, problem resolution, liability, responsibility, etc.

Table 4.2 Continued

Management Information	Inadequate, inaccurate, incomplete or untimely provision of information to support the management decision-making process	Inadequate reporting from all ICT systems
Ethics	Damage caused by unethical business practices, including those of associated business partners. issues include racial and religious discrimination, exploitation of child labor, pollution, environmental and so-called green issues, behavior to disadvantaged groups, sexual harassment, etc.	Inadequate content management to avoid offensive materials in e-mails, on web sites, etc.
Cultural	Failure to deal with cultural issues affecting employees, customers or other stakeholders; including language, religion, morality, dress codes and other community customs and practices	Lack of flexible language modules for all ICT applications
Climate, Weather, Environment and Geology	Loss or damage caused by unusual climate conditions, including drought, heat, flood, cold, ice, storm, winds or by geological instability such as earthquakes and subsidence	Inadequate site selection for data centers

Threat Scenarios

In order to gain a more detailed insight into the specific threats that you wish to consider in any given situation, the taxonomy of threats classification framework can be used to prompt ideas about threat scenarios. These scenarios provide a much richer set of information against which to make risk management decisions than you would get with a simple list of threats. Each scenario is described by several qualitative parameters. Table 4.3 shows the framework for describing a threat scenario. Table 4.4 to Table 4.7 contain supplementary information for this framework. All these tables are based on pragmatic principles rather than any theoretical framework and can be amended or extended as a result of further operational experience.

Table 4.3   Framework for describing a threat scenario

Parameter	Description	Example
Specific Threat	A threat selected from Table 4.2	Unauthorized code inserted into an application to Defraud the Organization
Threat Agent	A threat agent suggested by Table 4.1 but defined in more detail as in Table 4.4	Disaffected employee working in the systems development team
Capability	Level of resources expected to be under the control of the threat agent, as suggested by Table 4.5	Full skill set and tool set required for the task
Motivation	What motivates the threat agent, as suggested by Table 4.6	1. Personal Gain 2. Revenge
Opportunity	Description of the opportunity or level of access available to the threat agent	Full access to development code and development environment
Catalysts	Events or changes in circumstances that make the threat agent decide to act, as suggested by Table 4.7	Redundancy of employee
Inhibitors	Factors that may deter the threat agent from executing the threat, as suggested by Table 4.7	Fear of being detected, losing job and gaining a criminal record
Amplifiers	Factors that may encourage the threat agent to execute the threat, as suggested by Table 4.7	Belief that the rogue code can be hidden and covered and not attributed to an individual

Table 4.4   Threat agents

Threat Groups	Threat Agents
Natural Events	Wind
	Earthquake
	Flooding from rainwater, rivers, tidal surges, storms
Accidental Events	Fire
	Flooding from burst water pipes or tanks
	Explosions caused by malfunction of processes or services
	Structural collapse or damage from other assorted causes
Technical Failures	Equipment failure from fair wear and tear
	Equipment failure from poor design or implementation
Individuals	Human errors made by our employees
	Human errors made by employees of other organizations
	Human errors made by members of the public
	Gross negligence by our employees, past, present and future
	Malicious actions by disaffected employees
	Malicious actions by individuals belonging to external third-party organizations
External Organizations	Organized crime syndicates
	Terrorist groups
	State-sponsored action groups
	Competitive commercial organizations
	Political pressure groups

Table 4.5   Threat agent capabilities

Capability	Explanation
Finance	Money to finance the threat activities
Technical Equipment	Computers, specialized networking equipment, etc.
Software	Software tools to perform detailed analysis, probing and penetration of systems
Facilities	Buildings, services and general support
Expertise	People who are educated and trained in the techniques to be applied in mounting the threat activities
Experience	People with previous experience of mounting the threat activities

Table 4.6   Threat agent motivations

Class	Motivation
Personal Gain	Financial gain
	Revenge
	Gaining knowledge or information
	Exerting power
	Gaining peer recognition and respect
	Satisfying curiosity
Group Gain	Furthering the aims of political groups
	Furthering the aims of criminal organizations
	Furthering the aims of religious organizations

Table 4.7   Catalysts, inhibitors and amplifiers

Inhibitors	Catalysts	Amplifiers
Fear of capture	External events that trigger a Response	Peer pressure
Fear of failure	Changes in personalcircumstances creating a'need'	Fame
Inadequate access limiting the opportunity	Step changes in level of access increasing the opportunity	Easy access providing high level of opportunity
High level of technicaldifficulty	Step changes in level ofdifficulty through newtechnologies and tools	Ease of execution because of low level of technical difficulty
High cost of participation	Steep changes in level of cost	Low cost of participation
Sensitivity to adverse publicopinion	Dramatic changes in publicopinion and cultural values	Belief in sympathetic publicopinion

4.7.4 Risk Prioritization

It is not possible to mitigate all the risks all the time. What you need to know is which risks are the most dangerous and hence on which ones you should spend your resources on, for mitigation and management. The main reason for carrying out a risk assessment is establishing this ranking. The objective is to come up with a ranked list of risks showing the order of priority.

4.8 SABSA® Risk Assessment Method

The SABSA® approach to risk assessment is to adopt a qualitative measurement method that classifies risks into a series of bands. The following steps describe the method.

4.8.1 SABSA Risk Assessment Method: Step 1

Business Drivers and Business Assets

The assets are those things of value to your business that you wish to protect and maintain. The SABSA approach uses the business drivers and Business Assets concepts to capture the whole notion of assets. These are then turned into requirements for security.

4.8.2 SABSA Risk Assessment Method: Step 2

Threat Assessment

The SABSA method takes the view that since a detailed threat assessment is, often, too difficult for a normal organization to undertake, the approach should be binary, does this threat affect you yes/no? Then, you make a list of threats and threat scenarios that you consider to be relevant for your business. The Business Requirements from Step 1 are used to help frame the statement of a threat that will prevent that requirement from being met.

4.8.3 SABSA Risk Assessment Method: Step 3

Impact Assessment

Once the business requirements and the threats are stated, the next step is to assess what would be the business impact that would result from each threat materializing. This is first stated descriptively and then rated on a simple qualitative scale.

A qualitative scale might sound a little fluffy, but there is, currently, no simple way of quantifying concretely an impact. You could use monetary value for some risks, but others will have derivative consequences, like the loss of face, how do you quantify that?

4.8.4 SABSA Risk Assessment Method: Step 4

Vulnerability Assessment

Here the trick is to ignore any additional controls that you have already applied and focus on the vulnerabilities as if nothing has been done to mitigate them. This is often difficult, because people will want to explain what they have already done to mitigate the risk, but it is a major component in developing the security architecture to assess the impact of the planned controls before any security is being built.

The aim here is to state some control objectives at the conceptual layer of the security architecture and use them to develop and drive the detailed design of the logical, physical, component and operational security architecture levels.

4.8.5 SABSA Risk Assessment Method: Step 5

Risk Category

In the SABSA method, the prioritization of risk is based on 4 risk categories. These are calculated directly from the impact and vulnerability

Table 4.8   Risk Classifications

Category	Color Code	Description	Required Actions
A	Red	Severe Risk	Immediate action to mitigate and reduce the level of risk or reduce the impact level
B	Yellow	Significant Risk	Corrective actions should be planned and executed to reduce the vulnerability or the impact level
C	Green	Acceptable Risk	These risks are acceptable or are part of normal business in the industry
D	Blue	Negligible Risk	No Action

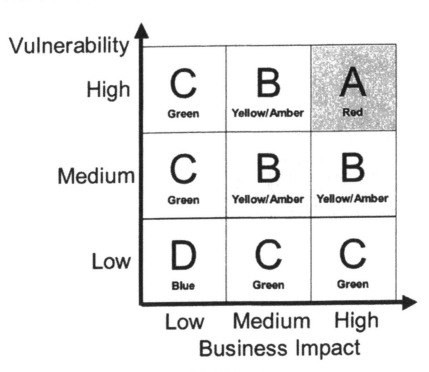

Figure 4.2   Risk Matrix

rating. Figure 4.1 and Table 4.9 provide more detail on the meaning of each category.

4.9 Business Processes and their Security

This part is relevant to the cell in the SABSA matrix, titled Business Process Model on the contextual row and in the process column (Chapter 3, Table 3.3). In addition to this, it provides the background for developing the deliverable title Business Process Model. This section looks at some of the sub-process elements that have more generic security requirements. This will help us identify the security requirements for the individual business processes.

4.9.1 Business Interactions

Interactions between the business entities have some of the following security requirements:

1. Identification of each business entity and their interactions with other entities
2. Authentication between the entities
3. Entity authorization restricts the actions to those that have been authorized.

The entities involved in the interactions may be:

1. Individual human entities
2. Corporate entities (entire businesses or business divisions or departments);
3. Logical entities (such as applications, acting on behalf of individuals or corporates).

4.9.2 Business Communications

Communication is an important part of many business processes. The following checklist reminds you of many of the methods used.

- Telephone conference calls

1. Mobile telephones
2. Video conferencing
3. Local area networks
4. Wide area networks.

Some applications of these communications methods include:

1. Home banking
2. Corporate office banking
3. Home shopping
4. Internet and web access
5. E-mail
6. On-line chat
7. Corporate networking and distributed applications
8. On-line transaction processing.

Each method of communicating comes with its own threats and vulnerabilities, and each application has its own business impacts. In developing the Business Risk Model, it is essential to consider the ways in which business processes are implemented and the types of technology that they employ to create the various systems.

4.9.3 Business Transactions

Use of on-line communications to transact business may include the following types of electronic transactions:

1. Contract's negotiation and agreement
2. Distribution of catalogues of goods and services
3. Specifications
4. Orders
5. Invoices
6. Payments
7. Transfers of ownership
8. Information delivery

Your own list will contain many more transaction types, often specific to your own business. In developing your Business Risk Model, you will need to examine each transaction type in context – what are the assets at risk, from what threats, what business impacts could result and what are the potential vulnerabilities?

4.10 Organization and Relationships Impacting Business Security Needs

This section is relevant to the cell of the SABSA® Matrix entitled Business Organization and Relationships on the Contextual row and in the People column (see Chapter 3, Figure 3.3). It also provides background for developing the deliverable entitled Organization and Relationships Model.

Some of the aspects of Business Organization and Relationships that you will need to examine to derive the business drivers and requirements for security include:

1. Management hierarchies and their effect on authorization, governance and control
2. Integrating the supply chain – trusted interactions between suppliers and customers

3. Outsourcing ICT operations to a third-party service provider
4. Strategic partnerships – how close you get, how much information you share,
5. Joint ventures – how much information is shared and how much is segregated
6. Mergers, acquisitions and divestments – whether the security architecture easily support changes in the overall structure of the enterprise (Once again the use of a security domain.)

4.11 Location Dependence

This section is relevant to the cell of the SABSA® Matrix entitled Business Geography on the Contextual row and in the Location column (see Chapter 3, Figure 3.3). It also provides background for developing the deliverable entitled Business Geography Model.

4.11.1 The Global Village Marketplace

The Internet has created a global village, almost a community, in which everyone from an individual, through small businesses, medium-sized businesses, right up to the largest businesses has access to the same marketplace, either as a customer or as a supplier. This has effectively removed many of the traditional barriers associated with the location. However, it has also introduced some interesting challenges for securing the business, since you can no longer see, feel and touch the other parties to your transactions. This distance and remoteness have a huge impact on the management of trust and will provide a key business driver for the Business Risk Model.

4.11.2 Remote Working

Another modern development fueled by the Internet is the trend towards workers no longer being in corporate offices, a trend that has accelerated with the recent Covid 19 pandemic. People can often work from home (teleworkers or telecommuters). Those who travel on business (the road warriors) can keep in touch using telephones and e-mail and have a virtual office that moves around with them, based on a laptop computer and a mobile telephone. This introduces a broad set of requirements to secure remote business information processing and communications in hostile environments connected over long-distance third-party networks.

Even those who do work in corporate offices will likely find themselves working in virtual teams, where the group consists of people in different countries and different time zones working as a team using information and communications technology as the means to communicate. Some of the team may be inside a corporate office somewhere in the world and others may be working from home, from a hotel room, from an airport lounge or from their car.

It is common for large organizations to have multi-site offices, often in different countries and for these offices to be linked by corporate networks supporting online communications between various parts of the business. In some cases, you will even find virtual companies that have no corporate offices at all where everyone is a home worker or road warrior, but these tend to be small hi-tech companies in the knowledge economy. All these potential modes of remote working are key business drivers for the Business Risk Model.

4.12 Time Dependency

This section is relevant to the cell of the SABSA Matrix entitled Business Time Dependencies on the Contextual row and in the Time column (see Chapter 3, Figure 3.3). It also provides background for developing the deliverable entitled Time Dependencies Model.

4.12.1 Time-Related Business Drivers

1. Business transaction turnaround times. The security architecture must support these
2. Business transaction lifetime. These affect the mechanisms that you apply to secure business transactions.
3. Business deadlines. For example, banking cut-offs and stock-market closing times may have an impact on how the security architecture is to be implemented
4. Record retention times. The security architecture needs to ensure that data can be retrieved and read and used right up to the end of the period, which is often a regulatory requirement.
5. Response to customers. It must be within a time that they expect. This is another example of the need to ensure that security mechanisms and procedures do not delay the business responses.

6. Just-in-time operations. These are needed for manufacturing operations where stock levels are kept to an absolute minimum to maximize cash flow in the business.
7. Time to market. This means balancing the risk of going to market with a product or service that may not be perfect in terms of its security against investing more time to get the security to an acceptable level whilst missing a business opportunity.

4.12.2 Time-Based Security

Consider an attack on a secure system. The time taken to break into the system depends upon how much security there is. No security, zero time. Some security takes a short time. High security takes a long time. It is an easy concept to understand.

Now, as soon as the attack begins there is a detection sub-system that starts to work. It takes a finite time for that sub-system to detect and notify management that an attack is happening. Once notified, it then takes management a further finite time to react to the alarm and to repel the attack. The effectiveness of the security or the exposure, can be calculated in terms of time, being a mathematical relationship between these parameters. The outline of the mathematics is shown below:

Time-Based Security

P is the Protection Value measured in time (= the time that the system will resist attack – the time it takes to break into it)

D is the Detection Value measured in time (= the time it takes for the system to raise the alarm)

R is the Reaction Value measured in time (= the time it takes for the system management to react to an alarm)

IF P > D + R THEN the system is secure and the attack will fail

IF P < D + R THEN the system is exposed and the attack will succeed

E is the Exposure Value measured in time (= the time during which the system is exposed and the attack can cause damage)

E = D + R – P.

The point of all this is that by measuring these time-based parameters you can begin to design systems that are more secure and less exposed. The aim is to reduce both D and R as close to zero as possible and to use this to select components in the component security architecture.

4.13 To Summarize: Contextual Security Architecture

Information security is a great enabler of business activities. It allows you to create solutions to business problems and to mitigate business risks to a level that is acceptable, such that these otherwise risky business activities can be carried out safely. The description of your business needs for information security is called contextual security architecture.

Information security is especially applicable in all types of digital business where the application of information and communications technology is used to create new ways of doing business. Specific applications include electronic publishing, on-demand entertainment, value-added information services, remote process control, supply chain management, research and information gathering, online banking, procurement and Government.

Information security is essential to maintaining operational continuity and stability in a modern business. The key dependencies include revenue generation, customer service, market reputation, management control, qualifying for operating licenses in regulated industries and maintaining employee confidence and shareholder confidence.

In safety-critical business systems, information security contributes to their safe operation by protecting remote communications against accidental corruption or malicious attack. It also contributes to providing assurance of the correct operation of such systems.

There are many business goals and success factors that are protected by information security architecture and many operational risks that are mitigated. The most important of these goals include brand protection, fraud prevention, loss prevention, business continuity management, strategic business development, fulfilling legal obligations and maintaining the confidence of key stakeholders.

5

Conceptual Security Architecture

The conceptual security architecture is where we as security architects begin to add value. At this stage, we have gathered and analyzed all the information collected about the business of the organization. What is needed now is a vision of the future, a concept of the types of solutions that will satisfy the business needs. If we get this part right, then everything else will flow with ease. If we get this part wrong, nothing at the lower layers of the architecture will ever be able to fix it. This chapter describes some of these key areas where we will need to apply our conceptual creativity.

In this chapter you will learn about:

1. The importance of conceptual thinking
2. How to develop the SABSA Business Attributes Profile
3. How to extend the SABSA Business Risk Model
4. How to use architectural layering techniques and how to apply these to multiple situations
5. A layered architectural model of security infrastructure and how to determine which services are best placed at which layers
6. Some of the major security strategies that you will need to include in our conceptual security architecture
7. The concept of a security entity and how entity relationships are characterized by the amount of trust between the parties
8. A method of analyzing complex trust relationships into simple components
9. The concept of a security domain and how this concept can be used as a powerful modelling tool
10. Some important lifetimes and deadlines that affect security
11. How to assess the current state of your enterprise security architecture as a basis for planning quick wins.

5.1 Conceptual Thinking

There is a skill called conceptual thinking that is important to becoming a successful architect. The key to mastering this skill is learning to stand back and being able to see the forest. Conceptual architecture is very much at the level of the forest. You are only concerned with the overall shape and size of the forest, the overall mix of tree species and the way they are to be grouped to create habitats for humans and wildlife alike.

Your concerns here are with the big picture, the helicopter view and the strategic plan for your security architecture. You must not be concerned with the details; these will come later, with a vengeance. The conceptual architect is a visionary, someone who can create a new vision of the future and sell that vision to others, leading the thinking of the architecture team and its clients. Think business here when you see clients! If this seems challenging, it is because it is, so let's get to some of the details and learning surrounding the conceptual security architecture.

5.2 Business Attributes Profile

Please refer to Chapter 3, Figure 3.3 the SABSA Matrix, Conceptual Layer, Asset's column, where you will see a cell entitled Business Attributes Profile. This part explains in detail how this Business Attributes Profile is used as a tool for conceptualizing the business assets that need protection in an information security architecture.

These Business Attributes have been compiled from the extensive experience of the SABSA organization and the architects there. The experience reflects work done with numerous clients in many countries and different industries. Over the course of that work, it has become clear that although every business is unique, there are many commonly recurring themes. This same experience is the background and foundation for lots of other frameworks, like CIS 18 or the Cloud Controls Matrix (CCM) from the cloud security alliance. This experience has been used to create a taxonomy of Business Attributes.

During the contextual and conceptual security architecture phases, the Business Attributes Database is used in two different ways:

1. To prompt your thinking on business strategies, business drivers, business assets, goals, and objectives
2. To map Business Attributes to business drivers in the Business Risk Model, using the risk assessment methodology defined in detail in Chapter 4.

The Business Attribute Profile is the complete set of Business Attributes that represents your business, mapped to business drivers and business risks and with measurements producing metrics and specific performance targets defined for each one.

This profile is a powerful tool that allows any business to be translated into common terminology and normalized. The profile selects only those Business Attributes that apply to this specific business. The taxonomy provides a checklist of possible attributes. The Business Analysts can decide whether a given attribute should be included in this specific profile. The senior executives should sign off on the overall Business Attributes Profile.

The Business Attributes Profile is an important conceptualization of the real business and forms a core part of the conceptual security architecture. It appears in the first cell of the second row of the SABSA§Matrix shown in Table 3.3 in Chapter 3.

It also allows the selection of metrics that are used to set performance targets as an integral part of the Business Attributes Profile that can later be measured to answer the question: 'Did we hit the target'? This too is at the choice of the business analysts, using either the suggested measurement approaches in the detailed definitions of the attributes or creating new measurement approaches if this seems more appropriate. Once again, the performance targets usually need to be signed off at senior executive level.

Thus, the Manage and Measure activity in the SABSA§Lifecycle is based upon the Business Attributes Profile that was set out during the Strategy and Concept phase of activity and which has been customized specifically to conceptualize this unique business.

5.3 Control Objectives

A control objective is a statement of a desired result or purpose to be achieved by implementing controls within a particular business activity. Controls are implemented through policies, organizational structures, processes, practices and procedures and through technical systems.

A control objective can be stated in response to specific business requirements for control or it can be a generic 'good practice' statement that should be applied to all businesses. This latter use of control objectives is at the heart of the COBIT 2019 Framework, from ISACA which focuses on generic ICT control objectives.

The SABSA Methodology uses control objectives to conceptualize the mitigation strategy developed through the Business Risk Model. During the contextual security architecture phase, you build a Business Risk Model. Like the Business Attributes, the control objectives are a way to take a unique real business and normalize it into common terminology and concepts that can be used to drive the more detailed design work.

You now decide upon the control objectives that best express your needs for security and control and insert them into column 11 of the Business Risk Model. This is an important interface between the description of the real business and the description of a conceptual model of the business. Columns 12 and 13 of the Business Risk Model are used to record the target vulnerability after the planned risk mitigation has been applied and the new overall risk category that results from that new reduced vulnerability level.

In selecting your control objectives, you can either create your own (and you will probably need to do this at least some of the time) or you can draw upon other sources of standard control objectives such as:

1. ISO/IEC 17799: 'A Code of Practice for Information Security Management'2;
2. ISO/IEC 21827: 'Systems Security Engineering Capability Maturity Model'3;
3. COBIT 2019: 'Control Objectives for Information and related Technology'4;

5.4 Security Strategies and Architectural Layering

Please refer to Chapter 3, Figure 3.3 the SABSA§Matrix – Conceptual Layer, Process column, where you will see a cell entitled Security Strategies and Architectural Layering.

There are many security strategies that you can adopt and many ways in which you can layer your security architecture. This section examines some of these possibilities in some detail, at a conceptual level. There is one word of caution needed for those readers who regard themselves as software architects.

This section has almost nothing to do with software architecture. Where it does address that concept, it is made explicitly clear. Thus when you examine the various layered models of security architecture, you must resist the temptation to translate them into software designs – they are not software designs. This chapter is about conceptual security architecture, not software architecture, and these layered models are conceptual models, not detailed logical designs.

Figure 5.1   Each layer in SABSA is the foundation for the layers above.

5.4.1 Multi-Layered Security

People often refer to the onion-skin model of security, where layer upon layer of defense is built up one on top of another. That analogy has been transformed here to simplify the diagram. The concept is shown in Figure 5.1. The figure represents the information assets that you wish to protect. Around that are multiple layers of security, each at a different level of detail. Closest to the assets are security controls that act directly on the information assets, cryptographic controls. As you move outwards the controls become more and more generic, until at the outer layer you have responsibilities, organization and policy.

The primary reason for a multi-layered mitigation strategy is to ensure that there is no single point of failure in the security measures. If one measure fails to stop a security incident, then there are others that do the job in a different way. The multiple layers provide a reasonable level of assurance that there are multiple ways of preventing security breaches. This is a fundamental principle that is strongly recommended that you adopt in your security architecture.

5.4.2 Multi-Tiered Incident Handling

Another ways of improving the effectiveness of your security are the provision of multi-tiered security services for dealing with security incidents. First,

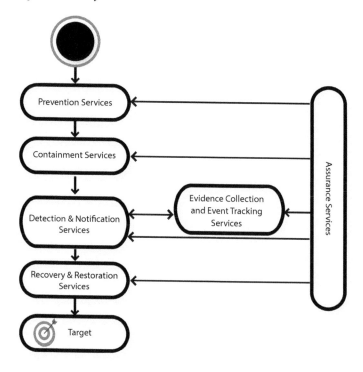

Figure 5.2   The individual services are key to reach the final target.

you try to prevent them. If that fails, you need to contain the effects. You also need to detect an incident and raise the alarm, then react to the incident to recover and restore to business as usual. You also need to collect evidence to track events and assist with restoration and use for forensic purposes. In addition, the entire process needs a level of assurance that it all works correctly. Figure 5.2 shows this multitier approach in diagrammatic format.

A full list of security services classified under these broad tier-headings is contained in the description of the logical security architecture in Chapter 6. In creating your security architecture, you should aim to have a mix of security services that provide adequate coverage in each of these tiers of this conceptual model.

5.4.3 Security Infrastructure Layered Architecture

The provision of security requires some security infrastructure. This should comprise:

1. Common security services delivered to applications through a common applications security
2. Security middleware to integrate and deliver the common security services across distributed applications (important with cloud security)!
3. Security services on platforms
4. Security services embedded in the network.

It is quite possible that in many organizations both the platforms and the networks may be outsourced for operations by a third-party service provider. Thus, they must be treated as separate security domains under the control of separate security policy authorities with separate security policies, so that no disruption or major operational difficulty is encountered at the time of outsourcing.

5.4.4 The Common Security Services API Architecture

In the real world, hardware and software are often provided as off-the-shelf components by vendors. The actual interfaces to these

devices are usually vendor-specific, although that is slowly changing. To integrate these various products into your architecture you need to construct the enterprise common security services API as a series of layered APIs.

This layered API framework is a conceptual model only. It is not meant to imply any specific software development method. The developers can use this conceptual approach to drive their actual software design.

It is essential to realize that this model implies sound overall software architecture, too! The upper software layers of the model must be designed from the get-go to support the extensibility and functional substitution at the lower levels. This is simply good software architecture. The implementation of a common application security services API as described here will have the beneficial effect of preventing that sort of sloppy software design and introducing proper architectural principles governing the development of software.

A Common Security Services API is maintained in-house by your systems development team or, if you do not have an in-house systems development capability, by a contracted systems house. Having that standard interface allows all your in-house-developed or custom-developed applications to see the same API, whatever underlying products are chosen.

The underlying products can be drawn together from several different vendors, each with its own proprietary interfaces. The products can also be changed and replaced without the applications needing to be changed since all the integration is done within the API architecture. This architectural approach effectively decouples the applications from the underlying product APIs, preserves flexibility and limits development costs.

There remains the issue of third-party applications, which like the lower-level products have their own vendor-specific API. These can be integrated into the entire API architecture by constructing application adaptors for each one. Application adaptors are software modules that convert the calls from the third-party application APIs into the standard calls of the enterprise-common security services API. With this additional sub-layer, the entire range of applications (in-house and third-party) can be integrated with the entire range of common security services.

The common security services are also likely to be supported by vendor products (such as PKI, Azure AD, AD, Cisco ISE, etc.) and these can be integrated in the same way as the applications. The best way to look at these common security services is to view them as pseudo applications with their own APIs.

5.4.5 Application Security Services Architecture

The previous section discussed the delivery of common security services to applications through layered infrastructure architecture. The applications part of this requires a more detailed discussion.

If you examine the range of legacy applications in most organizations, you will find that each one has been independently designed, developed and implemented, in many places running on legacy hardware and software as well. Each is unique and this uniqueness shows up in the way these applications are secured. However, the problem goes much deeper than just security. What you find is that each application has its own database and its own registry of users and that these are very difficult to share with other applications.

This model was not planned, it just sort of happened this way because of independent, uncoordinated developments. It is this limiting situation that has driven many organizations to take on the concept of strategic architecture to ensure that there is an overall vision of how applications are built, and how they can be integrated with one another when the business needs such integration.

The digital revolution has also drawn particular attention to these legacy problems because the building of digital business systems usually means integrating a web-based front end with several legacy back-end systems.

To address the problems posed by the poor integration characteristics of the legacy applications, ICT architects are taking a much more strategic view. In this view, there is a common infrastructure that is shared by all applications. The most important part of this infrastructure is the provision of a central data repository shared between the applications, like a data warehouse.

Around this central data repository are several common services, again shared by applications because they are common. A good example of a common service is printing. The central repository is also surrounded by individual applications, each of which makes use of the central data repository and the common services. There are also common external interfaces, such as the web interface. This model harmonizes perfectly with the desire to identify and centralize a series of common application security services, they simply take their place alongside other common services that applications need.

5.4.6 Placing of Security Services in the Architecture Layers

In this infrastructure architecture model, the discussions have focused on the common security services provided to applications. However, there are other layers in this model where security service can and should be provided. These include:

1. Middleware security services
2. Data management security services – provided within the databases or middleware
3. Network security services
4. Platform security services
5. Cloud Security services.

The question now arises – which security services are appropriate at each of these layers, including the application layer? The next five sections address this question.

5.4.7 Security Services in the Applications Layer

The focus of application security is to address the question of who can do what within the application, and how much. Authorizations are created through some suitable management process where business users are granted privileges. These privileges include things like:

1. Application functions they can use
2. Application data they can read, update and create
3. Limits on application transactions they can make
4. Separation of duties on some sensitive transactions where a second person is also needed to authorize the transaction
5. Sometimes there is a context-based set of rules governing the location of the user for certain activities.

Authorizations are then enforced by the system through a logical access control service. As a front end to access control, you also need authentication to prove that the claimant really is the authorized party, preferably by using MFA.

As a back end to access control, you also need audit trails to tell you historically who did what and when. You also need tools for creating and editing the authorizations and reviewing the audit trail, security administration in other words.

Application security services can be summarized under the six As:

1. Authorization
2. Authentication
3. Access control
4. Audit
5. Administration
6. Application-to-application communications security.

Legacy applications and third-party vendor applications are often characterized by having their own unique, built-in access control system. Integration of these systems between applications is frequently a right nightmare. The holy grail of application security architecture is the ability to integrate all applications under a single system for the six, using the integrated API approach described above.

Some major benefits of a wholly integrated architecture are:

1. Single window for user administration
2. Single authorization database
3. Single sign-on for the users.

The major architectural approach for providing this integrated, a single administration window, single authorization database, single sign-on model is through role-based access control (RBAC). This is discussed in more detail below in the section entitled Authentication, Authorization and Audit Strategy.

Application-to-application communications need some further explanation. The security services needed are:

1. Confidentiality
2. Integrity
3. Authenticity
4. Non-repudiation.

5.4.8 Security Services in the Middleware Layer

The function of middleware is to provide transparency in common services for distributed applications. Specifically, client and server applications do not need to know the details of each other's locations because the middleware handles all that stuff transparently and provides the application with a logical, service view of life. The location and distribution of the servers do not matter to the application and are hidden within the middleware, especially with the proliferation of cloud services, this is important. The objectives for middleware are to enable:

1. Seamless interactions between application components via a set of common APIs
2. Node, service and data location transparency
3. Scalability and extensibility
4. Reliability and availability
5. Vendor, platform, operating system and networking protocol independence.

Middleware commonly deals with the following types of basic services:

1. Remote procedure calls (RPCs) from client to server
2. Inter-process messaging management
3. Data management
4. Load balancing between physical servers for logical services
5. Inter-process resource sharing
6. Prioritization of application services
7. Security services management.

There are two approaches to providing security services within the middleware layer:

5.4.8.1 Explicit Security Services

Explicitly requested by applications through explicit security API calls. In this case, the application is aware of the security service and of the results of any security events. The application makes requests and gets reports back again. This is necessary to meet certain types of business requirements, for example, where an application needs to store its own audit trail of digitally signed transactions for evidence purposes and where the signature keys used to belong to and are in the explicit control of the application users.

5.4.8.2 Implicit Security Services

Provided from within the middleware transparently without the knowledge of the application. These are provided using resources (such as encryption and authentication keys) that belong to the middleware itself rather than to the application and its users. These services are needed to provide adequate security within the middleware infrastructure.

For example, when the middleware finds a server on the same physical platform, it will not be necessary to encrypt the inter-process communications. However, when a remote procedure call or Web API call, is made to another physical server, the request and response may need to be protected from eavesdroppers in their journey over the network. The objectives of providing security within the middleware are to:

1. Provide a secure infrastructure upon which applications can run
2. Offer explicit security API calls to applications
3. Enforce logical and physical security domains and domain policies
4. Protect itself from logical attack
5. Be capable of creating a trusted operating environment for entities that have established trust relationships.

Middleware security services, both explicit and implicit must be provided completely independently of any security services provided in lower layers, such as in the network layer. This is because it is dangerous to rely upon the existence of security services in another layer that is beyond the control of this layer and which cannot even be monitored by this layer to ensure that the security services are available and switched on. This view also harmonizes with the principles explained for providing security in the network.

Explicit security services in the middleware layer include all those listed as security services for the applications layer above and are called through the enterprise common application security services API. These services are discussed in greater detail in Chapter 6, Logical Security Architecture.

Implicit middleware security services include the following:

1. Entity authentication for entities making use of the middleware infrastructure
2. Entity authorization and role management
3. Logical domain access control based upon entity roles
4. Physical middleware node-to-node mutual authentication
5. Physical middleware node-to-node confidentiality of transmitted data
6. Physical middleware node-to-node protection of message and object integrity
7. Traffic flow confidentiality
8. Real-time security monitoring, intrusion detection and reporting.

At times there will be some limitations on the ability to implement this range of implicit security services due to the lack of functionality in vendor middleware products. There are also limitations that arise through performance constraints and the need to support legacy applications. However, the intention here is to help you to specify a target security architecture, which will not necessarily be capable of full implementation on day 1.

One of the important prerequisites for middleware security services is inter-operability across multiple platforms, which requires either middleware security service standards to be defined and adopted, and where there are multiple standards supported, the provision of suitable translation services from one standardized environment to another.

5.4.9 Data Management Security Services

Among the basic service types that are listed above in describing the function of middleware is data management. This has special requirements for security.

Data management has a dual role: It provides both access to the application information and protection of these information resources. The key to achieving these apparently contradictory goals is authorization, and the provision of appropriate security services within this layer is critical to the success of the overall application's architecture. The data management function embraces the following components:

1. Metadata management
2. Relational database management
3. Management systems
4. Database access

5. Data warehousing
6. Data mining
7. Transaction processing monitoring.

Thus, the important security services required in the data management sub-systems are:

1. Access control to data at the object level, using labelling mechanisms within the metadata as a means to match data classification to subject access privileges
2. Authorizations based upon business
3. Data availability protection, using a variety of backup and restoration techniques.
4. Data integrity protection within databases, to maintain a high level of confidence in the quality
5. Data confidentiality protection, ensuring that stored data is only revealed to authorized subjects.
6. Authentication of SQL requests and responses.

The important security management services required in the data management sub-system include:

1. The process for data classification
2. The designation of system owner roles and the execution of these roles
3. The use of standard naming conventions for data objects as a part of an integrated data architecture
4. The support for standard data formats to provide inter-operability with other organizations (such as support for XML13 formats).

5.4.10 Security Services in the Network Layer

This is where a lot of security has been applied traditionally. It comprises several sub-layers: a sub-net (OSI layers 1 and 2,) a network layer (OSI layer 3) and a transport layer (OSI layer 4). There is also a need to provide network management, which is addressed from a security perspective collectively with systems management. Network topologies include:

1. Local area networks (LAN)
2. Campus area networks
3. Metropolitan area networks (MAN)
4. Wide area networks
5. The Internet

6. Intranets
7. Extranets.

The goals for security in the network layer are:

1. To provide high-quality, highly reliable and highly available connectivity
2. To protect these reliability, quality and availability attributes
3. This includes the protection of the network management flows (DNS, ICMP, SNMP, etc.)
4. To prevent theft of bandwidth by unauthorized users.

The network layer security does not exist to protect the confidentiality, integrity, authenticity, or non-reputability of higher layer protocol data units, including middleware objects and messages and application layer messages. It should be a fundamental principle of your enterprise security architecture that these higher layers will provide their own protection for confidentiality, integrity, authenticity or non-repudiation.

The provisioning of application security within the network layer will be architecturally unsound because it locks application security into network technology dependence and it can never be truly end-to-end. When network technologies change, the application security is put at risk.

This is a controversial area in which there is widespread misunderstanding of how network and application security works, especially amongst the vendor community. Many vendors of network security products try to sell these products on the basis that they will protect application data.

There is some merit in providing a transparent confidentiality service in the network (in the form of a virtual private network – VPN), but this only affords protection against the real outsiders in the external domain. It does little to protect the confidentiality of information inside the enterprise domain. When one examines the range of security incident surveys that are published, from all sources, from all countries, on every occasion, these surveys agree that a vast amount of security incidents arise from inside the enterprise. Thus, focusing protection only against the malicious opponent from the outside is inefficient.

When you move from a pure confidentiality service to integrity, authenticity and nonrepudiation services it is impossible to provide these services to applications by embedding the services in the network. Few people understand this but consider the authentication of transactions between applications. If this were to be attempted in the network layers the following problems will result:

1. If there is an authentication failure, there is no mechanism available to report this to the application, yet it is the application that needs to know that this event has happened and it is a decision to be taken by the application as to how to handle this
2. Applications need to authenticate specific application data structures such as whole transactions. If authentication is carried out down in the network layer there is no guarantee of a one-to-one mapping between an application protocol data unit and a network protocol data unit.
3. Applications usually store audit trails as evidence of business transactions successfully completed. If digital signatures are applied and if there is a contractual liability associated with the use of this signature, then it is essential to store not only the message but also the sign that is associated with it.

What this tells you is that technologies such as SSL, TLS and IPsec have limited uses for application security. They do have uses, but they are not the panacea for solving security requirements that the vendors would have you believe.

Even when you apply encryption mechanisms in the network layers to provide confidentiality of application-level data, there is still the problem that the application has no visibility of whether the encryption is turned on. It is a leap of faith that the transmitted data is being protected by encryption. There have been many instances where network encryption gets turned off and application data continues to be transmitted in clear.

5.4.11 Security Services for the Information Processing Layer

This layer is concerned with the architecture and standards of the processing platforms, operating system services and peripherals.

Platform and peripheral types include:

1. Personal computers
2. Printers and plotters
3. Various I/O devices – document scanners, digital cameras, . . .
4. Data network devices – switches, hubs, routers
5. Telephones, IP Phones, mobile phones
6. Tablets
7. File servers, database servers, applications servers, mail servers
8. Storage devices
9. Mainframe hosts and servers.

On top of this, there is also a wide range of possible operating systems in use on these various hardware platforms. The strategic principles for providing security services in the information-processing layer are:

1. To reduce vulnerabilities in the information processing platforms and infrastructure
2. To segregate and isolate production platforms and environments from those used for development and testing
3. To provide and maintain highly trusted execution environments for highly sensitive data processing
4. To provide secure storage environments for highly sensitive non-volatile stored data.

The major security services to be provided at the information-processing layer are:

1. Physical security of the installation site to prevent theft, unauthorized physical access to the platform or malicious destruction
2. Environmental protection of the installation: electrical power protection, fire prevention, detection and quenching, flood prevention, structural stability, humidity and temperature control
3. Local user authentication with passwords and possibly smart cards or other tokens and possibly biometric devices
4. Local user access control, based on local authorizations, provided by the operating system
5. Local audit trails
6. Cryptographic services provided by local cryptographic sub-systems (hardware and software)...

5.4.12 Authentication, Authorization and Audit Strategy

The main components of an access control sub-system are:

1. The subject – a person or process
2. The object – the resource to which the subject is requesting access
3. The access control enforcement function
4. The access control decision function
5. The subject registry containing all the information on registered subjects
6. An access control list (ACL) associated with each object
7. The audit logging sub-system.

The decision is based upon a series of sequential questions:

1. Is the subject registered?
2. Has the registered subject been authenticated?
3. Does the subject's privilege profile contain authorization to access this object?
4. Does the access control list attached to the object authorize this subject to be granted access?

The traditional model of legacy computer systems is one in which each system or application has its own built-in logical access control sub-system. Thus, all the functions in the conceptual model above are implemented in each individual system. There are many problems with this approach:

1. Subjects that need access to many target systems need to be registered separately on each one
2. Subjects with multiple registrations may be given inconsistent access privileges on different systems
3. Subjects whose privileges need maintenance, as in the case of a change of job, need to have each and every registration updated
4. Subjects who leave the organization need to be deleted from every registry on every system, creating a risk that at least one registration will be missed.

To avoid these problems a strategic architectural approach is a must! This involves a centralized authentication service, like Microsoft AD or Azure AD for the cloud, and the implementation of role-based access control (RBAC). The main features and principles of this strategic approach are:

1. A single central authentication service is set up that authenticates all subject requests on behalf of all target systems.
2. Subject registrations are thus decoupled from target system registries and each subject is registered only once on the centralized authentication service registry.
3. A business analysis of all subject activity is carried out to define several subject roles. These might easily be mapped onto job functions and job descriptions.
4. Each subject is allocated one or more roles, and these are stored in the centralized subject registry.
5. Each target system is now set up to register roles rather than individual subject names in the access control lists (ACLs) associated with the objects in the system.

6. The precise mapping of a role onto a set of system objects depends upon the outcome of the analysis of business activities to determine what functions and data are needed to fulfil the business duties associated with a specific role.

The central access manager enforces the subject (user, process) to be confined within those decisions, meaning that unregistered, unauthenticated subjects are turned away. The target system also makes decisions:

1. Has the role been authorized to use the requested functions?
2. Has the role been authorized to make the requested type of access to the requested objects?

Based on these decisions, the target system also enforces the rules, disallowing any unauthorized requests the business benefits of role-based access control are:

1. Static roles definitions, which require little maintenance.
2. Ease of administration of users and their privileges, since these are concentrated into a single central registry and privileges database
3. Low administration overhead
4. Single sign-on for users because they interact with a central authentication service
5. Stable security policy because a role effectively defines a logical security domain
6. Stable, auditable configurations at target application servers
7. Improved control over the joining, moving and leaving of subjects.

Using the RBAC approach, the entire access control system is now highly distributed across many platforms and systems so some security mechanisms are required to protect the interactions between these physically separated parts of the overall system. This protection is achieved by applying cryptographic protocols. These can be based upon the facilities of a PKI or they can be entirely based on symmetric cryptography (as in the case of Kerberos, which is just such a protocol). These protection mechanisms are described more fully in Chapter 7 (Physical Security Architecture) and Chapter 8 (Component Security Architecture.)

Another aspect of the RBAC approach is decoupling of the user-to-machine interface and the machine-to-machine interfaces. Many of the traditional authentication and access control systems have been built such that the user enters a password at a client workstation and this is then transmitted over the network to the target server where it is checked. The obvious vulnerability in this approach is that the passwords can be intercepted and stolen as they travel through the network.

To avoid this vulnerability, the transmission of the password is replaced by a cryptographic authentication exchange protocol between the client workstation and the central access control server, like active directory. A similar protocol is used between the CAM and the target system, provided that the encryption algorithm is robust, any interception of the password will no longer be vulnerability.

Further strengthening of the user-to-machine authentication process can be achieved by using mobile apps like Microsoft Authenticator or Cisco Duo. The key common strength of all these methods lies in the cryptographic protocol replacing the clear text password transmission. The use of tokens together with a password or PIN is often known as MFA or Multi-factor authentication because you need to have both to succeed.

Another possible approach to improving the user-to-machine authentication is to use biometrics, like Windows Hello or a fingerprint. A more detailed discussion of these various mechanisms is in Chapter 7 and chapter 8 where the physical and component security architectures are described.

5.4.13 Security Service Management Strategy

There are two issues to be addressed here:

1. The management of security services
2. The security of service management.

The management of security services includes:

1. Provisioning of security parameters and privileges for users
2. Provisioning of security parameters for application systems
3. Provisioning of security parameters for embedded systems in equipment such as routers
4. Routine security operations to maintain the corporate systems in a state of compliance with security policy and standards
5. Security monitoring and intrusion detection to detect security incidents and collect information relevant to the problem management process
6. Security incident and problem management to recover and restore secure operations following a security incident.

The security of service management includes:

1. Authorization of operator entities that will perform service management functions.

2. Segregation of critical duties to protect the corporate information systems environment from the malicious actions of any single individual working alone.
3. Local and remote authentication of operator entities
4. Access control to service management applications
5. Secure service management protocols
6. Independent monitoring and audit of security operations management
7. Integration with existing service management infrastructure and organization.

The IT Infrastructure Library (ITIL) was created by the UK Government. It is a definitive reference source on ICT service management that has been widely adopted across Europe, Asia and Australasia. One of the ITIL publications is specifically dedicated to security management.

5.4.14 System Assurance Strategy

System assurance is concerned with the correctness, reliability, and proper operation of a system. There are several strategic areas of control that help you to provide the required level of assurance:

1. Control over systems development
2. Control over production systems operations
3. Software integrity protection and anti-virus controls
4. Content filtering to keep out unauthorized and illegal data
5. Protecting the integrity of mobile code
6. Functional testing
7. Penetration testing
8. Security auditing.

Each of these is discussed in detail in Chapter 12 since they are part of the ongoing operational security architecture. For constructing systems with high levels of assurance, such as in the case of safety-critical systems, there are a few additional tools and approaches needed. These include:

1. Redundant devices
2. Fault-tolerant architectures
3. Formal methods of specification and proof
4. Probabilistic risk assessment and fault-tree analysis
5. System modelling using finite state machine models
6. Tamper-resistance to defend against malicious attack
7. Human factors analysis, looking at the user interface.

You should also refer to Chapter 4 for the section on Safety Critical Systems and within that to the sub-section on Systems Assurance, where some of the key requirements are discussed.

5.4.15 Directory Services Strategy

Directory Service is one of several common security services needed within the layered security infrastructure architecture. This is because the directory service is a critical piece of infrastructure without which it is difficult to deliver many other security services. Just think how your infrastructure would be managed if you did not have Active Directory? It is the centralized repository of much security-related information about system objects (users, groups, machines ...).

In the context of describing directory services the word, 'object' has a very broad meaning. It includes all classes of objects, including user objects. Where the term 'subject' is used in other parts of the chapter, in this section it is referred to as a user object.

The main functions of the directory service in supporting other security services are:

1. Holding registered details of all objects of all object classes in the form of a distinguished name plus a variety of attributes.
2. The directory attributes of objects include all location and contact details, credentials, roles, privileges, certificates, authentication values, status information, state variables, cryptographic keys and so on.

5.4.16 Directory Services Strategy: Management

The directory service needs to be subject to strict access control, so that directory users gain access only to the subset for which they are authorized. Secure directory access methods require user authentication and cryptographic protection of the data exchanges.

The integrity and availability of the directory service must also be protected, almost at all costs, since without the directory service almost no other service can remain operable, this includes any cloud services we have integrated with our directory.

Thus, the directory service integrity and availability are of enormous business criticality. Suitable directory security management must be applied. This may include:

1. Physical access control to the directory servers and their location

2. Environmental protection to prevent fire, flood and structural instability
3. Sophisticated service management and monitoring of the directory and its service availability
4. Logical access control with a severely limited set of privileged users
5. Strong user authentication for privileged directory administrators (MFA)
6. A highly resilient directory service physical infrastructure.

There are several important management issues that must be addressed in directory architectures:

1. The directory service must be inter-operable with other directory infrastructures
2. The overall directory service may need to integrate many existing legacy directories
3. The directory must be able to limit transitivity of trust and inheritance of privileges to avoid the problem of uncontrolled inheritance
4. The directory service architecture must support specific performance characteristics such as a high read-to-write ratio in the directory enquiries.
5. The directory service architecture must be capable of running the service in a highly distributed environment.
6. The directory service must be almost infinitely extensible and scalable to accommodate future growth.

5.4.17 Directory Services Strategy: Objects

The directory service must support both entity objects and file system objects. These two primary types of objects will be subject to separate but related security policies.

1. Entity objects represent abstractions of such entities as users, roles, groups and hardware.
2. File system objects present the standard hierarchical file system
3. Access to both types of objects must be controllable by a variety of attributes and support for controlled inheritance.

Entity classes might include:

1. Organizational entities
2. User entities
3. Application entities
4. Hardware equipment or platform entities
5. Site and building entities.

All object relationships in the directory must be controllable by a hierarchical directory schema. All object classes must support multiple properties. The properties store and provide information about objects. For example: in the case of a user, this includes information such as:

1. Distinguished name
2. Home directory
3. Telephone number(s)
4. E-mail address
5. Roles.

All rights granted in the directory should be created via a standard template for that type of object. That means that when you assign security to a directory object, that object acquires the same rights as another object or group. If the rights assignment changes for one object (in this case the template) it also changes for all the equivalent objects. This greatly simplifies objects rights management and avoids the creation of inconsistencies.

You should avoid creating objects with partial equivalence or subsets of equivalence to represent partial overlaps in privileges. Users with requirements for different authorizations in different contexts will require multiple user objects to be created to represent them, like groups. For example, a user who is both a customer and an employee of the enterprise should be registered twice, once under each context. However, these objects must cross-reference one another to ensure that mutually exclusive authorizations are not allowed.

Objects in both X.500 and LDAP directories are defined as either container objects or leaf objects, depending upon their position in the hierarchical directory information tree.

1. Containers correspond to directories in a hierarchical file structure
2. Leaf objects correspond to files in the hierarchical directory
3. An object belongs to either a container or a leaf object, depending on the class to which it belongs, as defined by the directory schema
4. Names of objects are created from the complete path to the root starting at the leaf
5. Object relationships and access rights are based on a security equivalence list
6. To meet both security policy and architectural requirements, only security administrators should be allowed to create and delete directory objects in containers
7. System administrators can modify non-security-related object attributes in the systems under their purview.

5.5 Security Entity Model and Trust Framework

Please refer to Chapter 3, Figure 3.3, the SABSA Matrix, Conceptual Layer, People column, where you will see a cell entitled Security Entity Model and Trust Framework. This section discusses in detail some important conceptual models of entities and their trust relationships.

5.5.1 Security Entities

A security entity is something or someone that can take action in a business environment. These actions need to be controlled through authorization process and through technical and procedural controls that enforce those authorizations. Security entities are of several types:

1. Individual entities
2. Corporate entities, like an organization unit
3. Application or system entities.

5.5.2 Security Entity Naming

Each security entity must be identified with a globally unique name to ensure that there will never be confusion about which entity is being referenced, in Active Directory for instance a GUID is used as the unique identifier. However, local alias names are also permitted for use in the local domain where the alias is unique within the local domain but not necessarily unique within the global domain.

The directory is the repository for holding information on all security entities, including their globally unique name, any alias names and all other attributes including security attributes, as discussed earlier in this chapter under the heading of Directory Services Strategy.

5.5.3 Security Entity Relationships

Security entity relationships are characterized by the information flows that represent the relationship. There are three major types of entity relationship that you must consider:

1. Unilateral relationships – in which one entity broadcasts or publishes information and other entities may receive
2. Bilateral relationships – in which two entities make a specific contract to transact business and exchange information
3. Multilateral relationships – in which several entities participate in a group relationship.

Each of these security entity relationships implies a certain degree of trust (which is discussed in detail below).

5.5.4 Understanding and Modelling Trust

Consider a simple business model. A merchant has goods to sell and advertises them on the web. A web interface allows customers to browse the site, look at the goods, select what they want, place an order and make an electronic payment by sending a digitally signed authorization message to take money out of their bank account.

What needs to be protected here? What is the security that you require and what function does it serve? The answer can be summed up in a single word, TRUST.

The key security-related issue with all relationships between business entities is trust. In the simple merchant-customer business model the trust is implicit in the relationship between the seller and the buyer. It is also two-way trust, where both must trust each other. Now examine some of the ways that trust characterizes this relationship. The buyer must trust the merchant to:

1. Offer goods that are of the expected quality for the given price
2. Send the ordered goods, once the payment has been made
3. Not repudiate the receipt of a payment that has been received
4. Accept the return of the goods and refund the money if the goods fail to meet the expectation
5. Handle after-sales complaints about the failure of the goods to live up to pre-sales claims made about durability, fitness for purpose and so on.

The merchant must trust the buyer to:

1. Pay for the goods and have enough money in his bank account to cover the price
2. Not make vexatious or false claims about the quality of the goods
3. Not repudiate receipt of the goods that have been delivered
4. Not repudiate the order that was placed by the customer
5. Not repudiate the payment authorization.

This list is not exhaustive, but it shows you that trust is a multi-faceted and complex thing and that trust flows in both directions in many business relationships. It is essential to understand that trust is an attribute of relationships between business entities and that trust is not a technical attribute.

It is also quite clear that the types of trust and the levels of trust vary enormously from one business transaction to another. Each business relationship is unique in this respect. For example, if you buy a hot dog from a stall at a street market, you might wonder if it is fit to eat. However, you will not worry about whether the stallholder really owns the hot dog and has the right to sell it, nor will you question whether he will be there tomorrow in case you have a complaint to make. At the other extreme, if you are buying a house then you will employ a lawyer to investigate every detail of the ownership, the local planning regulations and the prospects for future peaceful undisturbed residence in the house once you have purchased it trust is not homogenous across all business relationships.

So, in a business transaction, who is the customer for trust? The answer to this question will give you an insight into how a trusted third party might sell trust services as a business opportunity. That is, you will understand the value proposition and business opportunity for a trusted broker, like a certificate authority (CA) providing guarantees for public key certificates (PKI).

The answer to this question will be unique to the specific trust relationship, closely related to the kind of transaction needed so consider again the simple earlier example, with a merchant selling goods over the web. Some very specific aspects of the trust model are used here to illustrate the point.

Take for example the issue of quality of goods. Who is the customer for trust in this respect? Obviously, it is the buyer of the goods. He is the one who must rely on the claims made by the merchant regarding quality.

As another example, who is the customer for trust with respect to payment for the goods? Here it is the merchant who must trust that the payment will be honored and will not be repudiated by the buyer. The merchant must rely on the electronic payment turning into real cash.

A pattern is being developed here, there are parties who make claims and other parties who must rely on those claims. This can be conceptualized into a simple model of a claimant and a relying party. The relying party is the trusting party who trusts the claimant. The claimant is the trusted party, who is trusted by the relying party. Most analysis will result in trust relationships which will boil down to this simple one-way trust model.

Of course, a business relationship between two entities will usually be characterized by complex two-way trust, which comprises several individual types of trust, each possibly at a different level. Each of these types can be identified and possibly further analyzed until there is only an array of simple one-way trust models aggregated together.

5.5.5 Protecting Trust Relationships – Trust Brokers and PKI

Using this analysis technique, you can reduce everything to a series of simple one-way trust relationships, which makes life a lot easier. You will want to protect these trust relationships with technical and procedural solutions and some of these will include PKI-based solutions. However, PKI is not the answer to every problem and there are many other ways to protect trust that in some circumstances will be more appropriate to the business need, like CASB (Cloud Access Security Broker) for cloud deployments for instance.

Part of the process of architecting the digital business infrastructure is to understand where PKI is the right solution and, equally important, where it is not. However, for the moment the focus remains on PKI to carry through the ideas from the previous section of this chapter and to see how it can be used to protect this simple trust model to which all other complex trust models22 can be reduced.

Clearly, it is the relying party that needs to purchase trust services. The claimant has only a passing interest in these services in that he wants to be trusted, but it is without doubt, the relying party who must be convinced that the trust is real. This means that the relying party must have a business relationship with the trust broker and must enter a contract with the trust broker by which the trust broker agrees to provide certain trust brokering services and the relying party agrees to pay for them and trust the services provided.

To provide a high level of assurance regarding the trust service being offered, the trust broker must also offer a remedy if the business transaction goes wrong. The trust broker must guarantee that he will take responsibility for the trust. So, trust broker services provide the wheels for digital business transactions, but when the wheels start coming off, the trust broker must be responsible and liable, otherwise, the service will fail to provide value.

In the late 1970s and early 1980s, when the concept of public-key cryptography was being discussed in technical textbooks on cryptography, it was assumed that the business opportunity for running a certification service would be in selling digital certificates. However, when you perform this rigorous analysis of how trust relationships really work, it becomes clear that this original assumption was deeply flawed. Selling certificates is not where the business opportunity lies.

A trusted broker provides the certificate to the claimant, but the customer for his trust services is the relying party. If he makes a contract with the claimant and sells him a service, it has minimal value and the real need for a trusted service remains unsatisfied, because the relying party got nothing at all and that is why the original technical publications had it all wrong.

We have already said that real business transactions are complex and involve two-way trust, such that both parties are simultaneously both claimant and relying upon party. But when you analyze this complexity into its constituent elements, there is only one element to be found, trust. So, whatever the business application, you must create trust broker services that meet the needs of the relying parties, recognizing that this will include all parties to a multi-party transaction.

5.5.6 Trust Broker Models that Work

The key to understanding the true business model is recognizing that what a trusted broker can sell is trusted transactions. The business model is like that already used for conventional credit card transactions. How do they make money then? Well, every time you use your credit card to buy something, they take a percentage of the transaction value from the merchant. Why would the merchant pay this percentage? Because the credit card company offers in return a guarantee that the merchant will get his money, whatever the status of the card, the cardholder or the account. If it is a stolen card and the merchant takes all reasonable steps to check for the authenticity of the cardholder, then the credit-card company takes the hit. The merchant is protected and so is the authorized cardholder.

What is on sale here is trust and liability management, which is exactly what you are looking for in digital business. Translate this plastic model into a digital model. The digital certificate is the equivalent of a plastic card. You can apply for one and you must be registered and pass certain verification tests about your identity, your creditworthiness and so on. If the registration authority is satisfied that it trusts you, you will be issued with a digital certificate by the CA. The relying party wants to know, 'Can I trust the claim being made by the claimant'? In some cases, it is simply a claim of a given identity.

In other cases, the claim will be that the claimant will behave under certain agreed rules and conditions, that he will pay his debts and that he will in general not cheat the relying party. The trust model here is what is known as *transitive trust*. It works like this:

1. The trust broker trusts the claimant. This trust is established through the registration process.

2. The relying party trusts the trust broker
3. Hence, by transitivity, the relying party trusts the claimant and trusts the claims that he makes.

This model also submits to the hierarchical analysis used earlier and that it can be treated as a triangular combination of three complex two-way trust relationships, each of which can be reduced to a series of simple one-way trust relationship models, as before.

5.5.7 Extended Trust Models for PKI

The model assumes that there is a single trust broker that has both registered the claimant and is known to and is highly trusted by the relying party. There are many cases where this will not happen. The claimant has registered with one trust broker that is unknown to or not well trusted by the relying party, and the relying party trusts a broker that has not been adopted by the claimant. After all, we all live and work in a free market where there will be many trust brokers with many differentiated offerings. In addition, some CA's might lose the trust of browser vendors, as we have seen in recent years, because of lacking processes for identifying the party that is buying a certificate.

One industry that has developed a good understanding of trust and trust brokerage is the banking industry. This is because banking is based entirely on trust between bank and customer. Who else do you trust enough to look after all your money and to give it back again when you ask? So, are these concepts being rolled out in banking? Yes, they are.

5.5.8 Levels of Trust

There is one more thing that needs to be discussed on the concept of trust, the potential need for different levels of trust for different business situations.

The level of trust that can be associated with a business transaction is directly related to the level of trust that the parties can have in the digital certificate that has been used to broker the transaction. This in turn is directly related to the strength of the registration process by which that certificate was granted and issued. How much validation was done regarding the identity of the claimant? If the processes a lax here, the CA will lose the trust of other CA's and the browser vendors, as mentioned earlier.

How much verification of trustworthiness was carried out? On the one hand, there might be situations where self-registration is the appropriate way for a given business application. You go to a public website, you fill in a form with your name and other details, you click OK and you are issued a certificate sent to your e-mail address. An improvement on this click-and-go process might be to check that the name and the e-mail address are related, but this may not work in all cases.

At the other end of the trust scale, you apply for a certificate in writing and you experience something like the process for getting your first passport. You need a birth certificate, probably a social security number and a Photo. Just to make certain, independent checks might also be made for a criminal record. Now that's serious registration!

As an aside, the use of birth certificates, passports, driver's licenses, banker's references, personal referees and the like are all examples of transitive trust, where an independent trusted third party is being used to verify some aspects of your identity or trustworthiness.

These very different needs for levels of trust and for appropriate registration processes to accompany them will lead to the provision of different classes of digital certificates designed for different purposes. The purpose for which the digital certificates are intended and the trust and liability management that can be associated with them are described in documents called the Certificate Policy (CP) and the Certificate Practices Statement (CPS).

5.6 Security Domain Model

Please refer to Chapter 3, Figure 3.3, the SABSA Matrix, Conceptual Layer, Location column, where you will see a cell entitled Security Domain Model. This section explains in detail the concept of security domains.

5.6.1 Security Domains

The security domain concept is a very powerful modelling tool. Here I present a definition of what it means and an explanation of the terminology, benefits and concepts used in that definition.

A security domain is a set of security elements subject to a common security policy defined and enforced by a single policy authority. The activities of a security domain involve one or more elements from that security domain, often with elements of other security domains.

A security element may be a security entity or a security object. A security policy expresses security requirements for a security domain in general terms. Security policy rules are derived from the security policy during engineering

Table 5.1 Assurance levels for registration processes

Registration Assurance Level	Implications for Trust
Self-registration	This allows very quick sign-up for customers on a click-and-go basis. It is suitable only for services publicly available within the user's environment (possibly a closed business environment) where the only service provider's interest is in collecting general information on how many and which users are registering for the service. There can be no attempt made here to ensure that the registered user is the authentic owner of the name claimed. However, this approach may be suitable for certain web-based services.
E-mail registration	This again allows quick sign-up by allowing the user to be authenticated based upon the possession of an e-mail address with a given domain name embedded in it, indicating that the user belongs to an organization with that domain name. The level of authentication is weak and the use of the domain name credentials will not cover all situations, especially for staff with other e-mail addresses. However, for certain low-assurance applications, this may be suitable.
Web-based credit card registration	In this case, the user self-registers but proves more about his identity by supplying a valid credit card number. This is still a relatively weak registration process in which it is easy to supply a fraudulent card number (which may or may not exist), but for low-assurance applications where small payments are involved and a level of fraud can be tolerated, this may be a suitable method. Suitable applications will include low-value information services.
Telephone confirmation	Adding a telephone confirmation process can augment each of the previous three methods of registration (self-registration, e-mail registration and web-based credit card registration). The user is telephoned at a number obtained independently of the initial registration ('out of band') and certain details are verified in the conversation. This eliminates certain types of impersonation.
Postal Registration	In this case, the entity sends an application for registration through the post, enclosing documentary evidence of identity and membership of the given business community. The policy of the registration authority determines the strength of this process, specifying whether copies or original documents are required and how many documents of what type must be presented. In the USA, there is additional strength and implication to using postal registration. Use of the postal service to defraud constitutes mail fraud, which is a federal offence, therefore adding additional risk mitigation potential.
Personal Registration	This is where an individual must attend in person to a registration office and present credentials for verification. The registration process can be made as rigorous as is required for the business environment and is entirely up to the registration authority in determining the registration policy. High value transaction or order-execution systems will require strong registration processes. They will also require strong user authentication mechanisms, such as smart cards or biometrics
Transferred Registration	Existing client databases can be used as sources of registration data. In this case previous registration details obtained for another purpose are accepted as suitable and are transferred to this application. The strength of the registration is entirely dependent upon the strength of the original registration and upon the level of maintenance to the database to keep its details up to date.
Delegated registration	In this case, a registration authority registers a sub-registration authority that is delegated with the responsibility for registering users within its own domain. These delegated registration authorities are often known as local registration authorities (LRAs). There could potentially be several levels or tiers of delegation. The overall strength of the registration depends upon each and every one of the processes at each level in the hierarchy.

activities. The security policy rules interpret the security policy in terms that can be incorporated into security mechanisms, which will, in turn, deliver security services that are used to implement the security policy.

A security policy authority is responsible for setting and implementing the security policy within the domain. For example, in a domain of registered users, the registration authority is the domain security policy authority. The policy authority might be a system owner, that knows the system is processing PII, and therefore must deal with the EU GDPR regulation.

5.6.2 Inter-Domain Relationships

Two security domains are said to be isolated from each other if they have no data objects in common and no activities in common and therefore cannot interact, a rare occurrence in many organizations, due to all the cross-system integrations that are happening at an increasing rate.

Two security domains are said to be independent of each other if they have no data objects in common and the activities within each security domain are constrained only by their own security policies and the security policy authorities of each are not constrained to coordinate or harmonize their security policies. Two or more independent domains may enter into agreements to coordinate sharing of information among them.

5.6.3 Trust in Domains

An entity is said to be trusted for some type of activity in the domain, in the context of a security policy, if enforcement depends upon an expectation of the entity behaving in an appropriate way. The security policy defines which entities are trusted and for each trusted entity the policy defines the set of activities for which the entity is trusted. An entity trusted for one set of activities is not necessarily trusted for all activities in the domain.

The security policy may require a mechanism to detect misbehavior by a trusted entity. A trusted entity that can misbehave without the violation of policy being detected is said to be unconditionally trusted. A trusted entity that can violate the policy but not without the misbehavior being detected is said to be conditionally trusted.

Trust within the domain is not necessarily two-way and not necessarily transitive. Transitivity of trust is usually defined technically by the security policy but is equally dependent upon the relationship.

5.6.4 Secure Interaction Between Domains

To be able to exchange information between domains the domain policy authorities must agree with a set of security policy rules governing this interaction in a secure manner. These jointly agreed rules form part of the policy rules of each individual domain.

To implement these secure interaction rules the policy authorities will have to negotiate an agreement on a set of common security services and common security mechanisms as well as on the information items to be exchanged.

If the interacting domains are both sub-domains of the same overarching domain, then the overarching domain policy authority may impose the security rules or it may allow the sub-domains to negotiate their own set of secure interaction rules, depending upon the terms of the overarching domain's own security policy.

5.6.5 Security Associations

A security association is a set of shared information items and attributes that describe a relationship between two or more entities. These security associations control the provisioning of security services involving interaction between the two entities. A security association implies the existence of interaction rules and the maintenance of information regarding this interaction.

5.6.6 Logical Domains

A logical domain maps into groupings of logical entities, such as entities sharing an application, a business community or a privileged group of entities. Logical domain boundaries are usually protected by logical access control, typically role-based.

5.6.7 Physical Domains

A physical security domain maps onto groupings of physical entities. It is usually a site, a platform or a network. Physical network domain boundaries are usually protected by firewalls. Doors, locks, gates, guards and so on are usually used to protect physical site domain boundaries.

5.6.8 Multi-Domain Environments

A given environment can support multiple overlaid domains of different types. A distributed application is a single domain, for instance, but it will likely be spanning multiple physical domains as part of its operation, whereas a single laptop is a single physical domain, with several different application domains running.

5.6.9 Applying the Security Domain Concept

The application of these concepts to real-world situations requires some discipline of approach, it but provides a powerful modelling tool. The following guidelines will help you to make the best use of this tool:

1. Make sure that your specification for a given domain is clear and concrete
2. Make sure the definition of the domain boundary and the domain interfaces are clear so that there is no confusion as to which elements and activities are included or excluded
3. Make sure you identify the domain security authority for each domain and that this authority really does set and implement security policy governing the domain
4. Be prepared to consider sub-domains and overarching domains
5. Many real business environments involve multiple domains, often overlaid and overlapping. Although this complexity may seem daunting, the domain modelling approach gives you a method to help you unravel the complexity and reduce it.
6. When designing control points such as firewalls and access control systems, make sure you cover all the domain boundaries and all the interaction points
7. Do not forget that there is always a surrounding domain called 'the universe', typically the Internet, that is a super domain of all domains and which should be regarded as completely hostile.

5.7 VPN Concept

Virtual private networks (VPNs) are seen by some organizations as a way to provide a secure business communications environment, but there are some limitations to be considered.

A VPN uses point-to-point encryption within the network layer to provide a series of communication lines, typically using the Internet or MPLS, along which private business data can be transmitted without being readable by an eavesdropper.

There are some issues to consider here. If you construct secure communications between two domains without regulating the flow at each end of the line, then all that the pipe does is to connect the two domains to form a single domain.

This is perfectly acceptable if the security policy and the ownership of the remote domains are identical, but if they are not, then the secure pipe becomes a means by which one domain can attack the other. In this case, you still need the firewall to regulate and control the flow of data between the two domains connected by the secure line. You need a firewall at each end of the line because each domain must be responsible for setting and enforcing its own security policy.

Typically, a VPN is built by using embedded encryption and decryption in the firewalls that provide a secure interface to a hostile network such as the Internet. The standardized approach to this is to use the IPsec protocol, a secure version of the IP protocol that provides encryption and authentication within the IP packet level protocol.

However, the IPsec functionality can be embedded in any end-system and so may tunnel through the firewall right up to an application server, as HTTPS does. The downside of this approach is that the firewall can no longer monitor the contents of the packets because they are encrypted and so VPNs are regularly terminated at the firewall.

At the client end, if the client is outside the firewall, like at home, like we have all been doing during the Covid 19 pandemic, then the VPN client resides on the user's PC and the VPN runs all the way up to this platform.

5.7.1 Firewall Concept

A firewall is a security gateway that sits on the boundary between two network domains, or even between several subnets, enforcing the security policy of one of those domains and regulating the flow and types of network traffic into and out of that domain. Firewalls are aimed at preventing unauthorized traffic flows and detecting unauthorized attempts to penetrate the security boundary created around the protected domain

When it comes to securing data networking environments, especially those in which an internal corporate network is to be connected to an external hostile network such as the Internet, most organizations will choose to use a firewall. However, the limitations of firewalls seem to be poorly understood, considering some of the key issues.

Firstly, the Internet is so pervasive that it effectively surrounds your enterprise domain. Think of the private corporate domain as being a small island in a large ocean called the Internet. On top of this, all the various computing components (Laptop, tablet, mobile phone) are all outside of the internal network, at least some of the time. How can a firewall protect these devices against attack?

As soon as this concept is clearly understood, it becomes obvious that the firewall alone will do little because the firewall can only regulate traffic that is directed through it. If the boundary around the rest of the domain is leaky and if traffic flows into and out of the domain other than through the firewall, then the firewall is ineffective and may as well not be there. For example, if a member of staff uses a mobile phone as a wireless access point, then the firewall cannot control or see the traffic, this is where defense in depth or multi-layered security steps in to fill the gap.

The firewall itself must also be properly configured and maintained. Firewalls allow certain traffic and disallow other traffic, but the rules governing this will undoubtedly change over time and maintaining an effective set of rules in the firewall is an ongoing process. So, you must be clear about what you will allow and what you will block at the firewall. This is the firewall security policy and without a properly stated policy, the firewall will probably not offer the correct protection.

Even when you have the policy correct, you must continue to monitor and check that the policy is properly implemented and that the firewall is correctly configured according to this policy. Many people think that firewalls are essentially technical gadgets that work in isolation.

5.8 Security Lifetimes and Deadlines

Please refer to Chapter 3, Figure 3.3 the SABSA Matrix, conceptual layer, time column, where you will see a cell entitled Security Related Lifetimes and Deadlines. This section explains in detail the main lifetime and deadline concepts that you need to consider.

5.8.1 Registration Lifetimes

Each registered entity is registered for a fixed period, after which the registration expires and must be renewed, just like IP address registration with DHCP. This prevents the build-up of dormant registrations, which are not in use because the entity is no longer operational.

A typical registration lifetime would be in the range of 1 to 3 years. However, some registrations are much longer. An example of this is the registration of each individual citizen at birth and the assignment of a social security number that remains operative for the entire life of that individual.

5.8.2 Certification Lifetimes

A registered entity can be issued with a set of digital certificates with which to authenticate messages and exchange encryption keys. These digital certificates also have a fixed lifetime. The digital certificate lifetime should never exceed the registration lifetime because if it did, the certificate holder would be in possession of valid credentials when his, her or system registration had expired. All entities making use of certificates of other entities must check the expiry date as part of the certificate verification process.

5.8.3 Cryptographic Key Lifetimes

All cryptographic keys must have a fixed operational lifetime determined by an expiry date, to limit the time that they are exposed to possible cryptanalysis by opponents. Some keys may also be limited in exposure by the amount of usage that they are allowed. In this case, the keys should expire either on the reaching of the expiry date or on reaching the usage threshold, whichever comes first.

Digital signature keys should have two lifetimes: one for the period over which signatures can be made and the other for the period over which those signatures remain valid and can be verified in operational use. Archiving requirements will determine the signature verification lifetime.

The actual lifetimes to be used must be specified in the security policy of the security domain that owns the keys. Factors to keep in mind are the type of usage, the strength of the algorithm, the key length, the system architecture and other factors that influence the assessment of vulnerability to cryptanalysis.

Cryptographic keys might also need to be archived for the purposes of data recovery. Under those circumstances, a second lifetime should be specified, to ensure that the keys are kept in archived form for as long as is required to grant access to the data over its expected archive lifetime. For most business data this period will be several years and may potentially be indefinite. Keeping track of the

5.8.4 Policy Lifetimes

Security policies also have lifetimes. Over a period, security policy often decays in terms of its effectiveness and its appropriateness and there comes a point when it needs to be reworked. It is a bad idea to make frequent policy changes because this leads to confusion and undermines confidence in the policy, but there should be a policy review cycle, to keep the policies relevant with the ever-changing threat landscape!

5.8.5 Rule Lifetimes

It is important to distinguish between policies and rules. While the policy should stay relatively unchanged, the rules needed to implement that policy may need to be changed quite frequently to keep up with the rapid changes in technology and threats.

5.8.6 Password Lifetimes

All passwords should be subject to enforced expiry, as well as complexity rules, at the end of a specified lifetime. The actual lifetime to be employed is a matter of the security policy of the domain in which the password is to be used.

Passwords may be expired voluntarily before the end of their lifetime, but it is also essential to have a minimum lifetime before which the password cannot be expired. This value is usually selected to be 24 hours, but its actual selection is a matter for security policy in the domain. This minimum password lifetime prevents an uncooperative user from resetting new passwords immediately and thus returning to the previous value that should have been retired from use. Most identity management systems, like active directory come with tools to help manage and control the use and implementation of a password policy.

5.8.7 Stored Data Lifetimes

Every application or service that stores data should be analyzed to define the storage lifetime business requirements for the data, especially with GDPR from the EU is a significant risk, if we are in violation of this!

There may be several phases in the overall storage lifetime, beginning with memory-cached data, moving to online disk-stored data and finally offline archived data.

The archiving lifetimes should also be defined for backup copies of data. One of the most important aspects of managing long storage lifetimes of electronic data (years)! is maintaining the technologies and technical products that will be needed to retrieve the data. For example, magnetic tapes (yes, they still exist) become useless if the tape machines needed to read them are no longer supported. Physical degradation of storage media is also a problem that must be understood and managed because over time the quality of magnetic media will often decay to a point where the medium is no longer readable.

5.8.8 Data Secrecy Lifetimes

The secrecy of some data must be maintained for many years. The actual periods of time required are a matter of policy determined by the policy authority that governs the business area concerned.

For example, in many jurisdictions, privacy law applies to personal information stored and processed during the lifetime of the human subjects, but not once they are dead.

It is essential to plan technology solutions that can maintain these secrecy lifetimes without failing through the passage of time. This impacts upon the selection of cryptographic algorithms and key lengths, considering likely advances in technology over the intended lifetime where these advances might make cryptography easier to attack. This is an extremely difficult area that needs careful strategic planning and expert advice.

5.8.9 User Session Lifetimes

When a user session is opened following a login, there should be a maximum lifetime for which that session can stay open. This limits the risk of sessions being kept alive long after they are no longer needed, which leads to a waste of resources and can also result in abuse of the session by an unauthorized party. Session lifetimes should be specified in the order of a few hours, to match patterns of working hours for authorized personnel.

5.8.10 System Session Lifetimes

In some cases, it is wise to introduce a session concept even for continuous processes that never stop. This enhances performance in secure data communications. A session context is established, in the form of security associations between the communicating entities, so that fast, efficient, symmetric cryptographic techniques can be used. Among the important state variables in the security, context are the session keys, used for either authentication, confidentiality and to prevent replay attacks, where an attacker records a session and replays it to gain access. The keys and their lifetimes will prevent such an attack from succeeding.

The session keys must be renewed from time to time to limit their exposure to cryptanalytic attacks, which leads to the concept of a session lifetime. In general session lifetimes in these circumstances are expected to be in the order of 24 hours.

However, in dial-up connections, the session lifetime may be considerably shorter because the duration of the dial-up call itself is short and this determines the session boundary.

5.8.11 Response Time-Out

Actual time-out values are a matter for local domain policy and depend upon the circumstances of the request, especially the business environment. Who is waiting? What is the actual business activity? And how long will they be patient?

Care must also be exercised if transactions are timed-out at a client, but which have actually reached the server and been serviced but simply failed to be notified to the requestor. The server must monitor requests to ensure that an incoming request from a client is not a repeat of a previous request, assumed by the client to have been lost, timed out at the client, but serviced by the server. There are several mechanisms that can be used to implement this control.

5.8.12 Context-Based Access Control

Context-based access control uses a set of context parameters to make access control decisions in real-time. These context parameters include the time of day, day of the week, location of terminal and type of network connection. Thus, it is possible to construct access privilege rules that limit access to certain times of day and days of the week. Typically, access rights are granted during normal working hours but not outside of those hours. Conditional Access in Azure AD provides these services to M365 and Azure customers. Integration into other applications can extend this to 3rd party applications as well.

Context-based privilege rules can be associated with roles, either on a central basis or on a local basis. The actual selection of those rules is a security policy issue within each security domain.

5.9 To Summarize: Conceptual Security Architecture

The conceptual security architecture provides the big picture, the helicopter view and the strategic plan for your enterprise security architecture.

The business and everything about the business that can be an asset in need of protection against risks is conceptualized into a standardized, normalized form, the SABSA Business Attributes Profile.

The assets identified in the Business Attributes Profile are used to drive a risk assessment method that presents a prioritized view of the enterprise risks. This risk assessment is used to develop a set of control objectives that conceptualize the needs of the enterprise for mitigating the risks. Make sure you allocate enough time and resources to this activity! This is the foundation for all the layers and activities that follows.

Layering techniques are an important conceptual approach to developing enterprise security architectures. The strength and effectiveness of the enterprise security architecture are improved by adopting a strategy of multi-layered security services. The infrastructure architecture is also modelled as a series of layers, and security services are placed within these layers to provide the most appropriate combination of services.

Other major strategies for the enterprise security architecture are also described in some detail. These include a strategy for authentication, authorization and audit, built around role-based access control; a strategy for secure service management; a strategy for systems assurance; a directory services strategy and a public key infrastructure strategy.

At the conceptual security architecture level, people or organizational units and any technical elements that represent them are regarded as security entities and their interactions are conceptualized as entity relationships. The degree of trust that exists in such entity relationships drives the need for securing the communications between the entities.

There are various levels of trust, largely depending upon the degree to which the parties know one another. In a business environment, this knowledge is derived from a registration process, which for high levels of trust must be rigorous.

Trust between two entities can be either one-way or two-way and in some cases, a third-party entity intervenes, in which case the trust is transitive. Sometimes these third parties act as trust brokers. The concept of a trusted broker is essential to many business models.

However, complex the trust relationships become, they can always be analyzed into a series of component parts, all of which are simple one-way trust relationships. This method of analysis renders all trust relationships capable of being understood in detail and this detail will guide many decisions going forward.

The security domain is an important concept that is used to build up domain models of real businesses, providing a means to understand how different security policies can co-exist, governed by different policy authorities, and how these different security policies interact.

This domain modelling approach helps you to understand how networks and applications are intertwined with one another and how their security policies differ to achieve the appropriate continuous level of security across the enterprise. These domain models also allow the development of sensible conceptual approaches to providing security in business environments that make use of the Internet, including providing protection by the deployment of virtual private networks and firewalls.

The time dependency of security is conceptualized through a series of security-related lifetimes and deadlines and many of these concepts are briefly described.

During the development of the conceptual security architecture, there is an opportunity to document the existing state of the enterprise security architecture. This current status provides the baseline against which a series of 'quick win' projects can be planned and executed.

6

Logical Security Architecture

The logical security architecture develops more detail to flesh out the bones of the conceptual framework that we developed at the second layer of the security architecture model. The logical layer is largely concerned with the functional view of security, defining a comprehensive set of functional requirements. It does not at this stage pay attention to the security mechanisms that will be used to deliver those functions, those are part of the physical security architecture at the next layer down. In this chapter you will learn about:

1. An information architecture model that distinguishes between knowledge, information and data
2. The different needs for securing static information and dynamic information
3. How a security policy is used as a logical encapsulation of business requirements
4. The need to have security policies at different levels of granularity related to one another in a hierarchical security policy framework
5. The wide range of logical security services that are needed to implement security policies
6. How security services are grouped into the layers of the multi-tiered approach to the security already described in Chapter 5
7. How security services are integrated into a series of logical security architectures,
8. Detailed descriptions of the most deployed individual security services
9. How entities are arranged into a framework called a schema, along with the various attributes that describe the entities and how this logical schema is used to structure a directory service
10. The definition of security domains within the network, the middleware and the applications to manage successfully a set of security policies set by different security policy authorities that govern these domains

11. How security domains can also be used to achieve segregation of groups of entities within the extended enterprise
12. The security management activities involved in the security processing cycle.

6.1 Business Information Model

Please refer to Chapter 3, Figure 3.3, the SABSA Matrix, Logical layer, Asset's column, where you will see a cell entitled Business Information Model. It is not the task of a security architecture team to develop the Business Information Architecture and so you will see that this is assumed to be a pre-existing model.

6.1.1 Information Architecture

Information is the logical representation of real business. In Chapter 2 the idea of information architecture was introduced. You will find there a brief introduction to this concept.

In this chapter this idea is expanded further, to help you to understand fully the logical assets that you are working to protect. One distinction that you need to clarify is that between data, information and knowledge.

These three ideas are associated with different layers of the architecture model. Table 6.1 explains each one and their inter-relationships.

At the logical architecture layer, you have business information. Information has the following properties:

1. Information is a logical representation of something real. For example, a customer is represented in your information system by customer information. This information includes everything you need to know about the customer to do business with him or her.

Table 6.1   Abstraction levels of the real business

SABSA Layer	Abstraction Level	Explanation
Contextual	Business	The real business context
Conceptual	Business Knowledge	Information that has been given business value and related to business context through interpretive or reflective intellectual
Logical	Business Information	Data that has been transformed and structured to have business meaning and relevance through intelligent analysis and synthesis
Physical	Business Data	Physical Business Data Raw facts and quantities that form the inputs and outputs of business processes and that are processed and stored during process execution

2. Information is structured data, organized into fields, records, files, tables, databases.
3. Information structures are related to one another both in hierarchical and peer-to-peer relationships.
4. Information is time-related, the information or its historical context is important.
5. Information is independent of location, the information exists independently of the physical location of the underlying data.
6. The quality of information depends not only on the content of the underlying data but also on the structure used to present the information and the analytical tools applied.
7. The success of information is best measured in terms of the user experience of using it.
8. When one talks of information assets, these are secondary assets, representing the real primary business assets that you want to protect.

6.1.2 Static and Dynamic Information

Static information is that which does not move or change in the short term. Examples of static information include:

1. Master records and files
2. Executable code
3. Configuration information for systems and applications
4. Historical information, like audit trails, transaction records and historical message records.

Dynamic information changes and moves in the short term and might only have a short lifetime between being created and destroyed. Examples of dynamic information include:

1. Real-time free-format messages such as used in e-mail or chat
2. Real-time structured application messages such as database queries using SQL
3. Real-time transaction information
4. System and service management real-time information exchanges.

The protection of both static and dynamic information requires security services such as:

1. Confidentiality protection
2. Integrity protection
3. Availability protection.

The protection of dynamic information requires security services such as:

1. Authenticity of source
2. Non-repudiation.

However, many more of the Business Attributes can be applied directly to the protection of information.

6.2 Security Policies

Please refer to Chapter 3, Figure 3.3, the SABSA Matrix, Logical layer, Motivation column, where you will see a cell entitled Security Policies. Security policies and security policy architecture. Both deliverables are addressed in this section, although you will also need to read Chapter 9, Security Policy to complete your understanding of how to develop them. They are built on the concepts of security policy, security domain and security policy authority that was introduced as part of the conceptual security architecture in Chapter 6.

6.2.1 Security Policy: A Theoretical View

Security policies are statements of what type of security and how much should be applied to protect the business in various ways. Security policy is positioned at the logical layer of the security architecture model and is derived directly from drivers in the conceptual layer. The Business Attributes profile and the control objectives are strong drivers of security policy. So are aspects of the Security Entity Model and Trust Framework and the Security Domain Model.

A security policy defines what is meant by security within a security domain, the high-level rules for achieving this security and the activities that are to be authorized to achieve security objectives. The policy also defines how entities outside the domain can interact with entities inside the domain. For a definition and explanation of security domains please refer to Chapter 6.

The domain owner sets the security policy for the domain. The owner may delegate the implementation of the security policy to a lower security authority that acts on behalf of the domain owner. Such a delegated security authority is effectively the custodian of the domain. For a discussion on ownership and custody, please refer to Chapter 9, in the section Outsourcing Strategy and Policy Management'.

Thus, a security authority is an entity responsible for the implementation of a security policy on behalf of the owner and may also be the owner. The security authority may delegate the enforcement of a security policy to other entities within the security domain.

The security policy is determined by the business requirements for information management and information systems, following an assessment of the possible operational risks. Operational risk assessment is discussed in Chapter 9 and operational risk management is the subject of Chapter 10.

The security policy states what should be done but should as far as possible avoid any reference to technical solutions. For this reason, security requirements are expressed in terms of generic security services. Security services are discussed in detail later in this chapter.

6.2.2 Security Policy Architecture

Security policy exists at several different levels, it is useful to conceive of a layered security policy architecture. Figure 6.1 provides an example for such a layered architecture.

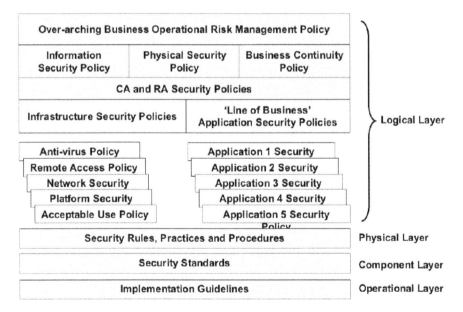

Figure 6.1 A Suggested Hierarchical Policy Architecture

You have probably seen similar layered diagrams before, but this is different because it does not have as its top layer the corporate information security policy, this appears on the second layer. The reason is that information security management is only one of multiple disciplines involved in corporate risk management and information security management needs to be closely integrated with several other related operational risk management disciplines.

So, there are policy statements that are applicable to all these related disciplines. It is better not to repeat these same statements under different policy headings but to assemble them together into an integrated, overarching top-level policy. Such a top-level policy is addressed to all employees throughout the enterprise.

There is a much more detailed discussion of this integration of operational risk management in Chapter 10, but for the present, you need to understand how this integration affects the structuring of the information security policy architecture. As a part of the logical security architecture, you will need to determine what is the appropriate policy architecture for your enterprise and which policies you will need at each level. Within this policy architecture, you will also need to populate it with the policies themselves, but the process for you to develop your detailed policies is described in Chapter 9, see also Chapter 3, Table 3.2, operational architecture at the logical layer. Figure 6.1 also shows several lower layers of documentation that support the policies, these too are discussed in detail in Chapter 9.

6.3 Security Services

Please refer to Chapter 3, the SABSA Matrix, Logical layer, Process column, where you will see a cell entitled Security Services. In this section, the most common security services are introduced and described.

The security services are logical services, specified independently of the physical mechanisms that might be used to deliver them. They are driven from the layer above, most specifically from the Business Attributes Profile, the control objectives and the security strategies.

6.3.1 Common Security Services and Their Descriptions

In Chapter 6 the concept of multi-tiered security was introduced and presented in Figure 6.2 as a layered model of security services including:

1. Prevention services
2. Containment services

3. Detection and notification services
4. Event collection and event tracking services
5. Recovery and restoration services
6. Assurance services.

This is now explored in greater depth by presenting a detailed list of security services under each of these six defensive strategy headings, see Table 6.2.

Table 6.2 Security services by defensive strategy

Defensive Strategy	Security Services

Prevention	Entity Security Services:
	  1. Entity unique naming
	  2. Entity registration
	  3. Entity public key certification
	  4. Entity credentials certification
	  5. Directory service
	  6. Entity authorization
	  7. Entity authentication
	  8. User authentication
	  9. Device authentication.
	Communications Security Services:
	  1. Session authentication
	  2. Message origin authentication
	  3. Message integrity protection
	  4. Message content confidentiality
	  5. Security measurement and metrics
	  6. Security administration (privilege management)
	  7. User support
	  8. Physical security services
	  9. Environmental security services
	  10. Non-repudiation
	  11. Message replay protection
	  12. Traffic flow confidentiality.
	Application and System Security Services:
	  1. Entity authorization
	  2. Logical access control
	  3. Audit trails
	  4. Stored data integrity protection
	  5. Stored data confidentiality
	  6. Software integrity protection
	  7. Software licensing management
	  8. System configuration protection
	  9. Data replication and backup
	  10. Software replication and backup
	  11. Trusted time
	  12. User interface for security.
	Security Management Services:
	  1. Security policy management
	  2. Security training and awareness
	  3. Security operations management
	  4. Security provisioning
	  5. Security monitoring
	  6. Security measurement and metrics
	  7. Security administration (privilege management)
	  8. User Support
	  9. Physical Security Devices
	  10. Environmental security services.

Table 6.2 Continued

Containment	1.	Entity authorization
	2.	Stored data confidentiality
	3.	Software integrity protection
	4.	Physical security
	5.	Environmental security
	6.	Security training and awareness.
Detection & Notification	1.	Message integrity protection
	2.	Stored data integrity protection
	3.	Security monitoring
	4.	Intrusion detection
	5.	Security alarm management
	6.	Security training and awareness
	7.	Security measurement and metrics.
Event Collection/Tracking	1.	Audit trails
	2.	Security operations management
	3.	Security monitoring
	4.	Security measurement and metrics.
Recovery & Restoration	1.	Incident response
	2.	Data replication and backup
	3.	Software replication and backup
	4.	Disaster recovery
	5.	Crisis management.
Assurance	1.	Audit trails
	2.	Security audit
	3.	Security monitoring
	4.	Security measurement and metrics.

In subsequent sections, each security service is described. You will need to decide which of these services you require in your enterprise security architecture to meet the requirements and policies that you have derived. However, no list of security services will ever be complete and you should not take this one to be so. It is a useful guide and is reasonably comprehensive.

6.4 Security Service Integration

A critical aspect of the logical security architecture is fitting these various security services together into a single integrated whole. This group of security services acts on or for security entities. Security entities are defined in Chapter 5.

6.4.1 Unique Naming

For each entity, there must be a unique name to prevent confusion over which entity is being referenced. The structure and syntax of this name will depend upon the type of entity. There must be a set of syntax rules and a service that creates and registers these names for new entities.

6.4.2 Registration

Each entity is registered as being part of the community of entities in the overall universal domain governed by your security architecture. Registration is an extremely important security control because if an entity can become registered, it can easily obtain privileges, perhaps even elevated privileges. The importance of the strength of the registration process has been discussed in Chapter 5.

6.4.3 Public Key Certification

An entity that has been registered likely needs to participate in interactions with other entities. If public-key cryptography is in use and PKI is included in the overall security architecture, then each registered entity requires a set of private keys and a set of matching public keys. There must therefore be a service by which the entity can generate the public and private key pairs and submit the public keys for certification. Public key certification prevents an unauthorized, unregistered entity from becoming a malicious participant in the business community being secured.

6.4.4 Credentials Certification

In a distributed systems environment, like the cloud, where a role-based access control strategy is adopted, the entity roles and other credentials information are sent across the network from the central access manager (CAM) to the target application server. The target server needs to trust the credentials contained in that information. It trusts the CAM, but it must also be assured that what it receives is the same that was issued by the CAM. This can be achieved by wrapping all the necessary credentials in a form of a certificate that has been digitally signed by the CAM security authority. These certificates are sometimes known as authorization certificates or privilege attribute certificates (PACs).

6.4.5 Directory Service

The directory service is built on four basic models:

1. The directory information model
2. The directory-naming model
3. The directory functional model
4. The directory security model.

6.4.6 Directory Service Information Model

This information model defines the types of data and units of information that you can store in your directory. This model defines the building blocks for your directory. The basic unit of information is a directory entry, which is a collection of information about an object. If you are familiar with Active Directory or LDAP, then the following classes of information will be familiar to you.

Objects are different types and each of these types is known as an object class. Some object classes refer to real-world physical objects such as people and devices. Others can refer to abstract objects such as roles. The following are possible examples of object classes:

1. objectClass: person
2. objectClass: role
3. objectClass: device
4. objectClass: site
5. objectClass: building and
6. objectClass: top.

Some object classes are sub-classes of a higher-level object class, just like with inheritance in object-oriented programming. Thus, the object class building is a sub-class of the object class site. There is a single top-level object superclass to which all other object classes belong as sub-classes. This root object class is called the top.

A directory entry comprises a set of attributes4. Each attribute describes one of the traits of the object being described. An object from the object class 'person'. Its directory entry (description) contains its 'distinguished name', the object classes to which it belongs and the set of attributes.

In the following example, the abbreviations used have the following meanings:

1. dn – distinguishedName
2. ou – organisationalUnit
3. uid – userIdentifi er
4. cn – commonName
5. sn – surname.

Each attribute has a type and one or more values, for example:

1. dn: uid=tmadsen, ou=asset_management, ou=people, ou=
2. objectClass: top
3. objectClass: person

4. cn: Tom Madsen
5. eMail: tma@somewhere.com
6. telephoneNumber: +1 123 456 7890.

Each attribute type also has syntax rules that control what types of data can be used for the values of that attribute. The syntax rules also contain information on how the directory matches values when searching:

1. For the syntax rule caseIgnoreString applied to the attribute sn, the values 'Madsen' and 'madsen are the same and would be matched.
2. For the syntax rule caseExactString applied to the attribute sn, the values 'Madsen' and 'madsen' are different and would not be matched.

6.4.7 Directory Service Naming Model

The distinguished name is the result of a naming model that arranges the objects in a hierarchical logical structure. The functional model of the directory describes the operations that you can perform on the directory to interrogate it, populate it and manage it. These functions are implemented through a set of protocols. The X.500 standard defines four different protocols, each with its own function.

By far the most popular of these protocols is Lightweight Directory Access Protocol (LDAP), which provides all directory access and management functions in a single protocol. LDAP is a client-server protocol. The server agent is in the directory server, often an AD server and every user of the service has the LDAP client software installed on the remote computer, often a laptop and the like, but it can equally be a database server with multiple users accounts for different DB services.

The functions provided by LDAP are arranged in groups:

1. query operations
2. Update operations
3. Authentication and control operations
4. Extended operations.

To make these functions work there is a sophisticated search engine on the LDAP server that can apply a variety of matching algorithms and filters.

6.4.8 Directory Service Security Model

The directory service is one of the most important security services. With the expanded use of the cloud, where the identities are distributed to different cloud vendors, this importance is only increasing. It provides a trusted repository for all entity information and is used by all other security services that need that information. Thus, being a security service, it requires integrated security services to maintain its own security and integrity.

The directory must be subject to sophisticated access control so that users are able to get access to entity credentials for which they are authorized. The directory access methods should require user authentication and cryptographic protection of the data exchanges.

The integrity and availability of the directory service must also be protected, at all costs I would say since without directory services no other service can remain operable. Suitable directory security management services must be applied:

1. Physical access control to the directory servers and their location
2. Environmental protection
3. Sophisticated service management and monitoring of the directory and its service availability
4. Logical access control with a severely limited set of privileged users and limited transitivity and inheritance to avoid the problem of uncontrolled inheritance
5. Strong user authentication for privileged directory administrators and operators.

The directory must also be inter-operable with other directory infrastructures, which implies conformance to industry standards such as LDAP or X.500, probably the first of these.

6.4.9 Authorization Services

Authorization services prevent unauthorized entities from gaining unauthorized access. There are three distinct parts to an authorization service.

The first part is an off-line service, in which a registered entity is granted privileges by a registration authority and those privileges are stored against the entity name as attributes of the entity in the directory. If role-based access control is being used, the privileges are in the form of authorized roles, the roles having been created in advance.

The second part of the service is the local offline administration of authorizations at target servers, associating specific local privileges with roles.

The third part is the real-time online authorization of a request made by an entity. The CAM first authorizes the request based on the role in the entity credentials and these credentials are then forwarded to the target application server. The target server uses the trusted role in the credentials to grant specific privileges to the entity based upon the role associations set up earlier in off-line mode.

6.4.10 Entity Authentication

Entity authentication means that one entity that claims a certain identity proves to the satisfaction of another entity that he or she really is the entity claimed. Entity authentication is a huge subject and a book of this wide scope cannot go into it in detail. For those with an interest in pursuing the subject further, there is a good reference in ISO/IEC 10181.

6.4.11 User Authentication

User authentication is a special case of the generalized entity authentication discussed previously. In this case, the claimant is a human user. You must consider the human-to-machine interaction as a special part of the overall exchange.

In Chapter 5 there is a discussion about the decoupling of a user from the network authentication exchanges, reducing the user interaction to a local exchange between the human user and the local PC. This is a case of indirect in-line authentication. The user is the claimant and the PC or terminal is the trusted intermediary.

The user has a password or a PIN or a biometric that is entered into the PC. This may be verified locally or may be used to construct the authentication information to be sent on in the second stage, like in the Kerberos protocol.

6.4.12 Communications Security Services

Communications security services protect the communications between remote entities from threats. The main threats are those perpetrated by an eavesdropper who listens in and possibly alters the message or mounts a denial-of-service attack.

When two parties establish a communications session between them there is a threat from an unauthorized eavesdropping third party that the session will be hijacked, either during the session setup or at some time during the session itself. Such a style of attack is often called a man-in-the middle attack. The opponent may completely take over from one of the participants and masquerade as that participant or may instead stay as a middleman altering the exchanges of information that pass between the original parties.

To minimize the vulnerability to this type of threat the session can be authenticated. The session setup is handled through a mutual authentication exchange that may or may not involve a trusted third party. Both parties can adopt the roles of both claimant and verifier, mutually verifying each other's identity.

To secure the session over its entire lifetime the exchanged authentication information should include some secret cryptographic keys not disclosed to the eavesdropper. Various secure key exchange mechanisms can be used for this (see Chapter 7, Security Mechanisms section). These keys are then used to authenticate every data exchange between the two parties throughout the session. This does not prevent the opponent from making changes to the exchanged data, but it does mean that any such changes will be detected.

TLS is a good example of session authentication with strong mechanisms, although in this case the authentication is usually limited to authenticating the server but rarely the client.

6.4.13 Message Origin Authentication

When a message is delivered to a recipient it usually contains information about its origin, which sent it. However, there is a threat that an unauthorized party can send a message pretending to be someone else. To minimize this vulnerability the message can be authenticated using a cryptographic mechanism (see Chapter 7, Security Mechanisms section). Each message is individually authenticated to allow the recipient to verify the claimed identifier in the message.

6.4.14 Message Integrity Protection

Even if a message originates from an authentic source, there is a threat that an eavesdropping third-party opponent can alter it in some way during its transport through the network. To minimize the vulnerability to this, attack the contents of the message can be authenticated by use of a cryptographic mechanism such as a message authentication code (MAC) or a message integrity checksum (MIC, see Chapter 7, security mechanisms section). Each message is individually authenticated for its contents. Any unauthorized changes will be detected, and appropriate action can be taken.

If the contents include the identifier of the originator, then that same mechanism can be used to provide both services, origin authentication and contents authentication. If the message contents include a sequence number that is held as a state variable12 at both sender and receiver, then an attack that attempts either to change the sequence of messages or to delete one of the messages in the sequence altogether can also be detected.

6.4.15 Message Replay Protection

There is a threat that an opponent may capture a transmitted message and replay the same message later, perhaps several times. By doing so the opponent could masquerade as an authorized user and conduct a faked login or he could alter the contents of payment messages to gain some financial advantage.

As a further extension of message integrity protection, if a nonce value is incorporated into the message contents it is possible to detect this attack and take appropriate action. A nonce value is a value that occurs only once and can never be repeated. Such one-time values are constructed from time stamps, random numbers, sequence numbers or combinations of these types of numbers. If timestamps are used, the time service must be a trusted service.

6.4.16 Non-Repudiation

When a message is sent by one party to a recipient it is often important for business reasons to prevent the sender from later attempting to deny that the message was sent or the recipient from later denying that the message was received. A non-repudiation service provides this assurance, either by use of an asymmetric cryptographic mechanism or by use of a trusted third-party logging mechanism (see Chapter 7, security mechanisms section). In the case of providing proof of receipt, a non-repudiable acknowledgment message needs to be created and sent.

6.4.17 Traffic Confidentiality

In some circumstances, it may be sufficient for an opponent to know simply that there is a peak volume of message traffic, even if all the messages are encrypted. By looking at the surrounding environment, the timing and the sources and destinations of the messages, the opponent may be able to draw some intelligent conclusions about what the messages are saying.

For example, in a military environment it could become apparent that an attack is about to be launched.

To minimize the vulnerability to this type of traffic analysis, a traffic flow confidentiality service hides the volume of messages. The mechanism used (see Chapter 7, Security Mechanisms section) is usually some type of dummy messaging to show a continuous smooth volume of traffic.

6.5 Application and System Security Services

This group of security services protects applications and systems from attack or abuse. They are mostly concerned with preventing or revealing unauthorized access or unauthorized actions by those who have been granted authorized access.

6.5.1 Authorization

The authorization has already been discussed under the heading of Entity Security Services above. You should read that section and this one in conjunction with one another since they are both concerned with the different aspects of authorization services.

Setting up the roles for an application service or set of resources is a policy matter for the security authority responsible for policy in that domain. The process requires a careful analysis of business needs to identify the roles that should be used. The aim should be to create a limited set of roles to minimize role access administration at target servers. In some applications, it may be acceptable to have only one role.

The business analysis of role requirements may include the creation of roles that are mutually exclusive for separation of duties purposes. Thus, a user entity may not possess both roles. It will be the responsibility of the application domain security authority to ensure that this policy is upheld in the granting of roles to requesting entities.

When a real-time access request for an application service is received, the CAM makes intelligent first-level access control decisions based upon:

1. The role of the requesting user
2. The static associations registered under the resource object to which access is being requested.

The real-time association of a role to an entity during an access request is dependent on several things:

1. The static role(s) stored in the directory as an attribute of that entity
2. Inheritance of role attributes under the directory schema
3. Pre-condition rules stored with a roles object in the directory:
4. Controlling transitivity within allowed limits
5. Configuring delegation controls
6. Checking and enforcing specific user conditions and constraints.

6.5.2 Access Control

Theoretical models for access control have been described at a conceptual level in Chapter 5. You may find it useful to refer to that section now.

Access control services control both physical and logical domains and access control mechanisms may also be physical or logical in their nature. There are three types of access control service:

1. Controlling access to physical domains such as sites and buildings using things like gates, doors, locks, guards
2. Controlling access to logical domains such as systems, applications, files, records and databases using logical access control mechanisms
3. Controlling access to physical domains such as hardware platforms and networks using logical access control mechanisms.

6.5.3 Audit Trails

Audit trails provide historical evidence of activity for monitoring purposes or forensic examination purposes. Protecting the integrity of the audit trail itself becomes an issue in many circumstances, since tampering with an audit trail may cover up unauthorized activity. Many compliance regimes require controls with logs. Thus, a robust audit trail service needs not only mechanisms for capture and storage of the event information but also mechanisms to protect the integrity of that stored information.

6.5.4 Stored Data Integrity Protection

Just as message data in transit can be subject to unauthorized alteration, deletion or resequencing by an eavesdropper, stored data can suffer the same fate between the time it is stored and the time it is retrieved for use. The security mechanisms for detecting such modification are the same as for transmitted data, using a MAC or a MIC (see above under Message Integrity Protection and Chapter 7, Security Mechanisms section).

The use of physical and logical access control mechanisms also helps to prevent unauthorized access that would lead to such unauthorized modification of the stored data.

6.5.5 Stored Data Confidentiality

This service prevents unauthorized disclosure of stored data. Several mechanisms are available (see Chapter 7, Security Mechanisms section) including encryption, physical enclosure of the data store and logical access control. Software integrity is a huge problem. The most significant threat comes from malicious software in the form of viruses, worms, macro viruses and Trojan horses.

Rogue software may also be inserted into a system manually by a hacker who has already penetrated to a level of high privilege and installs malicious code objects for future use, although most of those insertions are automated these days. The mechanisms used to implement the services to defend against malicious attacks include antivirus scanning tools, change-detection mechanisms such as checksums and quarantine environments for testing newly imported software before it is released.

It should be noted that it is impossible to prevent software from being attacked it should be viewed as normal for business systems. Your only real defense is to try to catch as many infectious agents as possible before they do too much damage and to be ready to clean up when inevitably some of those agents penetrate your defenses. There is no such thing as the silver bullet that kills all malicious code.

Software integrity-protection services also include the acquisition and distribution of third-party software packages to ensure that software is obtained from reputable sources and that it is clean of malicious infections when it is acquired. For organizations that develop their own software, either for internal use or for distribution to others, the process of releasing and publishing the software also needs to be controlled.

6.5.6 System Configuration Protection

The configuration of a system includes both the executable software, including scripts and the configuration data that many of the executable files need to perform their function. All these files and the directory structure in which they are stored need to be protected from unauthorized changes. This service is usually delivered by applying several security mechanisms including:

1. Anti-virus scanning
2. Use of checksums to check the integrity of files and directories
3. Use of scanning tools comparing the actual configuration with a stored configuration policy file.

6.5.7 Data Replication and Backup

To enable recovery of a system following a disaster incident the data must be backed up. This replication, backup and restoration service must cover:

1. The regular backup copying process
2. Backup media management: labelling, indexing, off-site secure storage, retrieval, encryption
3. Data restoration process
4. Backup and recovery sub-system testing.

6.5.8 Trusted Time

In distributed systems time is often used as a means of agreeing with certain aspects of a protocol or timestamps for log entries. One specific application of time is to include a timestamp into a protocol data unit to prevent message delay, message replay or message re-sequencing by an unauthorized eavesdropper. These timestamps are protected from alteration by using a cryptographic protection mechanism.

However, even if the opponent cannot tamper with the message itself, if he can tamper with your clock, he may still be able to persuade you to accept a message that is out of time because you no longer know what the time really is. So, the provision of trusted time service is a critical piece of security infrastructure.

6.6 Security Management Services

Security management services fall into two groups: procedural security management services and technical security management services. The possible list of such services is almost endless, but some key examples are included here.

The use of the term security policy, in this case, is a loose one. It means all the policies and standards that drive the configuration and management of the managed security objects. A security management agent is a software sub-system embedded in the managed end-system that handles the security management protocol and implements the instructions from the security management center. The security management protocol has primitives such as GET, SET and TRAP as found in most protocols for systems management and network management.

6.6.1 Security Measurement and Metrics

At the security management center, the data collected by the security monitoring service must be collated and analyzed to report management information, including measurement of performance in the form of agreed metrics. Some of the performance elements of the environment that might be measured include:

1. Security services response times
2. Preservation of security policy across domains
3. Confirmation that the authorization and authentication process is functioning properly
4. Confirmation that non-repudiation services are operating correctly
5. Relationship between the actual observed system behavior and standard security baselines for diagnostic and planning purposes.

Developing appropriate security metrics is important to the evolution of enterprise security architecture. The result of this activity will be to know with confidence whether the security management systems are working and how well. Several approaches to this endeavor are available:

1. The first is to create a reference system by which components of the production system are compared. This is accomplished by paralleling portions of the production system with the reference system and comparing functionality in terms of throughput and integrity.
2. Another testing method would include sampling line activity and subjecting the sample to FFA (Fast Fourier Analysis)16 to ensure encryption is functioning properly and to test for unexpected or unauthorized traffic.

Inspection tools, automated when possible, can be acquired or developed to test actual resource configurations against standards or expected configurations. Unfortunately, creating objective metrics for measuring the security of an organization or infrastructure is a difficult task. Make sure that the measurements you decide on can be monitored and are as objective as possible.

The interactions of numerous quasi-intelligent entities in a modern distributed computing network, with cloud components, tend to cause the environment to move towards the nature of a complex adaptive system and increasingly become subject to chaos, without good management practices. It may be necessary to assess at what point this is likely to occur and take steps to bound or limit the eventuality. Failure to anticipate this situation may lead to unwelcome results. Modelling and simulation tools are needed to address these issues.

6.6.2 Intrusion Detection

If a break-in or attempted break-in takes place, it must be detected as soon as possible and reported so that incident-response services can take appropriate action and recover from the attack. The service is implemented through the deployment of detection agents like IDS/IPS and the correlation, collation and analysis of this information at the security management center. Indicators of intrusion incidents can include:

1. Multiple instances of the same user
2. Failed logon attempts
3. Attempted access to unauthorized resources
4. Unusual network conditions
5. Components failing integrity tests
6. Unknown source addresses.

Detection of certain attack signatures by specialized intrusion-monitoring software. The agents are deployed to monitor both host platforms and network components.

6.6.3 Incident Response

Incident response services deliver actions in response to detected security incidents. In many cases, an incident or group of incidents will require a formal decision process, what should be done next? Who to contact? The decision about what action to take can in some cases be automated and in other cases will require human intervention. The logical steps required for an appropriate incident response include:

1. Data collection
2. Data normalization and collation
3. Data analysis

4. Incident assessment and conclusions
5. Presentation
6. Response alternatives
7. Response decision
8. Response action management.

The first action as the result of a reported security incident, once analyzed, is to determine whether an automated response is adequate or not. This decision can be made based on error type or a number of errors and can be controlled by scripts or logic engines. For example, an instance of a particular error type may be determined to be addressable by an automated response, whereas multiple instances may require human action. This could be an instance of a single resource failure as opposed to multiple resource failures.

The second decision involves any correlated error conditions. In most cases of system failure, it becomes progressive and may require a rapid response to contain the damage. A good approach is to develop worst-case scenarios and analyse your network and application resources so that a minimum sustainable configuration is known and can be quickly implemented.

This would include physical domain segmentation, human intervention, firewalls at critical domain boundaries, alternate communications links and other required resources.

6.6.4 User Support

Many operational problems experienced by users of systems and applications are security-related. The potential impacts of unresolved problems are lost production time and bringing the security services into disrepute within the business. There must be adequate user support services through the help desk function to manage these problems and ensure their timely resolution.

6.6.5 Disaster Recovery

Disaster recovery relies essentially either on organizational measures or on technical measures that have a broader scope than security. It is often regarded as part of business continuity management (see the section on BCM in Chapter 12). The mechanisms that support disaster recovery services include:

1. Taking appropriate backups of data and software
2. Providing backup management: labelling, indexing, storage
3. Off-site storage

4. Data recovery and restoration procedures
5. Redundancy of hardware and communication lines for resilient operations
6. Recovery plans and procedures
7. Contingency sites
8. Incident management responsibilities
9. Activation plans.

6.7 Entity Schema and Privilege Profiles

Please refer to Chapter 3, Figure 3.3, the SABSA Matrix, Logical layer, People column, where you will see a cell entitled Entity Schema and Privilege Profiles. In this section, the notion of a schema is explained and its application in structuring the security attributes of all enterprise entities is described.

6.7.1 Entity Schemas

A schema is a set of rules that determines what data can be stored in a database, directory or XML file. The purpose of the directory schema is to:

1. Help maintain the integrity and quality of the data stored in the directory
2. Reduce duplication of data
3. Impose constraints on the size, range and format of data objects stored in the directory
4. Provide a well-documented and predictable method for directory-enabled applications and services to access and modify the collection of directories
5. Help to slow down the effects of directory entropy, in which over a period with constant use by many entities, the contents of the directory tend to move towards less structure.

Before a directory server stores a new or modified entry, it checks the entry's contents against the schema rules. Whenever directory clients or servers compare two attribute values, they consult the schema to determine what comparison algorithm to use.

The components of a schema are:

1. Attributes
2. Attribute syntax rules
3. Object classes.

An attribute type definition includes:

1. A unique name identifying the attribute type
2. An object identifier that also uniquely identifies the attribute
3. An associated attribute syntax and set of syntax rules
4. A usage indicator (which applications use this attribute)
5. Restrictions on the range and size of the values that may be stored in the attribute.

An object class is used to group objects that have something in common, usually real-world objects of the same type, such as people, printers or network devices. A single directory entry describes an object that can belong to one or more object classes. Thus, a network printer belongs to the objects class 'printers' and to the object class 'network devices'. An object class definition includes:

1. A name that uniquely identifies that class
2. An object identifier that also uniquely identifies that class
3. A set of mandatory attributes that must be included in the entry describing the object
4. A set of optional attributes that may be included in the entry describing the object
5. An object kind (structural, auxiliary or abstract).

6.7.2 Role Association

The association of a role with an entity is achieved by defining for the entity object an attribute that carries the assigned role. Such an attribute might be called 'roleAssignment'. Since an entity may have more than one role, this attribute is often multi-valued, with more than one occurrence in the object entry:

1. dn: uid=tmadsen, ou=asset_management, ou=people, ou=architects
2. objectClass: top
3. objectClass: person
4. cn: Tom Madsen
5. roleAssignment: internetUser
6. roleAssignment: accountingClerk

In designing the schema, you need to define attributes for role management. Depending upon your requirements and your design decisions, the attribute may be mandatory or optional in the object class definition.

You will need to define the logical roles that are to be mapped to groups of users through attributes. This can only be achieved by a thorough business analysis of the job functions and how privileges need to be allocated to run the business. You should aim to minimize the number of roles whilst still maintaining sufficient granularity to achieve segregation of job types and their access privilege profiles. Some of the main role types may include:

1. User roles
2. Business manager roles
3. System manager roles
4. Operations management roles
5. Administrator roles
6. Auditor roles

Within each of these role groups, you will need levels of granularity depending upon your analysis of the business models. For example, in a financial accounting department, roles might be:

1. Financial director
2. Financial controller
3. Accounting supervisor
4. Accounting clerk
5. Financial auditor.

6.7.3 Authorization, Privilege Profiles and Credentials

Many aspects of authorization are discussed in other parts of this chapter. The discussion here focuses on the storage of authorizations in the form of privilege profiles, privilege attribute certificates or credentials. All these terms mean roughly the same thing.

If you are using role-based access control, as you should, then all that needs to be stored in the directory entry for the entity being granted authorization is the associated role or roles. This is achieved through defining attributes for the entity object, see above in the section Role Association.

For other types of access control management, where roles are not used, a similar approach can still be taken. Whatever forms the package of information that you call the entity's credentials, this information can be put into an attribute or set of attributes in the directory entry for the object.

Roles or credentials or whatever you call them are also objects. So, you might define an object class called 'role', which has a set of attributes that define a role.

6.7.4 Certificates and Tickets

Often a set of credentials is protected cryptographically, either to protect its integrity or its confidentiality or both. Such a structure is often a certificate (if an asymmetric cryptographic technique is applied) or a ticket (if a symmetric cryptographic technique is applied).

Certificates can sometimes be enduring data structures that could be stored in the directory as objects or as attributes of objects. For example, a digital certificate is usually stored as an attribute of the directory object representing the entity to which that certificate has been issued.

Some certificates and all tickets are transitory objects that have limited lifetimes, that is, they are temporary credentials. The most useful thing to do for these would be to define an object class in the directory and to store an enduring template of the structure used for creation.

6.8 Security Domain Definitions and Associations

Please refer to Chapter 3, Figure 3.3 the SABSA§Matrix, Logical layer, Location column, where you will see a cell entitled Security Domain Definitions and Associations. This section describes the most important types of security domains that you will need to define.

6.8.1 Network Domains

The logical specification of a network into domains, like subnets or segments, is a very useful step in developing your enterprise security architecture. The important thing to remember in doing this is the definition of a domain, a set of security elements that are all subject to the same security policy.

The typical enterprise has several different business units. Each line of business is potentially subject to different security policy restrictions regarding the sharing and disclosure of information.

Additionally, over time the shape and size of the business change because new businesses are acquired or existing businesses are sold. Some business units are not wholly-owned but are joint ventures with other companies. For all these reasons it is a good practice to design each business unit as a separate logical domain so that domains can be added or removed painlessly according to changes in business structure and so that potential differences in security policy between one business unit and another are easily implemented.

Note again the difference between logical and physical architectures. An extranet is a logical definition. At this stage, no assumptions are made about how that extranet will be physically implemented and there are several alternatives available, including the use of VPN technology, the Internet or all of these.

The organization also has a few people who regularly work from home and a few people who are on the road needing to communicate from laptops with mobile telephone connections, from hotel rooms, from business centers or from airport business lounges.

To facilitate this external access, another logical domain is created like a VPN connection or a cloud application, through which the application functionality and information that needs to be externalized can be delivered securely and safely to those externally located.

Additionally, both these groups of people are company employees and need to have access to a variety of information services and office automation support services that are bundled into a service package historically called intranet services but increasingly provided via the multiple cloud services offerings.

No one except the data center operations staff has direct hands-on access to production business applications and the intranet/extranet services. These applications and services are contained in an inner sanctum domain called production. Business users gain access through an inter-domain multilayered architecture that is conceptually like that described in Chapter 5.

6.8.2 Middleware Domains

The middleware is the services integration layer of the infrastructure. It provides transparency of location for application servers, application clients and common services used by applications. A very important component with the rising importance of cloud services to the modern enterprise.

In Chapter 5 there is a discussion of the types of common security services and the delivery of such services through a conceptual layer called middleware. At the logical architecture layer, this leads you towards a definition of a few service domains in the overall middleware domain.

6.8.3 Application Domains

Each application is a logical domain, subject to a security policy for that application. This application domain has sub-domains, which are best mapped onto the roles, each role being a logical sub-domain.

The real users are part of another domain, which you can call the 'people domain' and the information resources used by the application are part of an 'information domain'. There are mappings from these external domains into the application domain to associate people and information resources with the roles.

The extended application domain includes these external logical components. Thus, an extended application domain comprises:

1. Roles –the roles and functions and information associated with each role and the user to role mappings
2. Functions associated with each role
3. Users
4. High-privilege users
5. User groups
6. Information resources accessed by each role.

6.8.4 Security Service Management Domains

The security service management domain comprises:

1. The managed security objects and their management agents
2. The security management information database
3. The security service management functions (within security service management applications);
4. The security management personnel.

6.8.5 Policy Interactions Between Domains

The interactions between entities in different domains are controlled by the way in which the domain policies govern the interactions.

Take for example the situation shown in Figure 6.3. Domain 1 and Domain 2 are independent domains, but they are both sub-domains of Domain 3. Alice and Bob (A and B) are registered members of Domain 1, having been registered by the security authority SA1. They are subject to a security policy set by SA1. Similarly, Xavier and Yvonne (X and Y) are members of Domain 2,

having been registered by SA2 and subject to the security policy set by SA2. The authorities SA1 and SA2 are both registered by SA3, the security authority that governs Domain 3. SA3 sets the security policy for Domain 3, which in turn applies throughout both Domains 1 and 2. The subdomains have their own additional policies that supplement that handed down from the super domain.

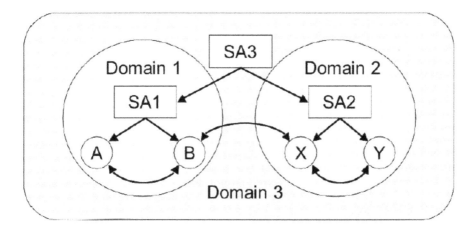

The policies of the two sub-domains cannot (by definition) conflict with the policy of the super-domain.

When Alice and Bob interact, they do so govern by the security policy of SA1. Similarly, Xavier and Yvonne interact governed by the security policy of SA2. However, when Bob and Xavier interact, they are governed essentially by the security policy of Domain 3, constrained by any additional policy requirements placed on them by both SA1 and SA2. They cannot behave as they might if only operating with their home domain.

6.9 Security Processing Cycle

Please refer to Chapter 3, Figure 3.3, the SABSA Matrix, Logical layer, Time column, where you will see a cell entitled Security Processing Cycle. This section expands on the meaning of this. The security processing cycle involves a few security management activities such as:

1. Introducing and registering new organizational entities
2. Introducing and registering new users
3. Setting up authorized privileges Registration renewal
4. Certificate issue and renewal
5. Provisioning and configuring equipment throughout the environment.

There are also several automated processes, such as for setting up and closing sessions, and for handling messages that have a defined time to live so that they are discarded if they prove to be undeliverable. To define the logical flow of each of these processes you will need to adopt a well-governed method. Here are some of the key considerations:

1. What is your complete list of security processes?
2. What event initiates each of these processes?
3. What event closes the process?
4. What intermediate stages are there in the process where it moves from one state to another?

What events trigger the transition of the process from one state to another? This is a relatively informal approach, but what is being described here is a loose version of finite state machine modelling. It can be applied here to each of your security processes to define their precise operational cycle.

6.10 To Summarize: Logical Security Architecture

Business information is the logical representation of real business. Hence the information assets needing protection are secondary assets that logically represent the primary asset, the business itself.

The overall logical representation of the business is through the information architecture, which must exist ahead of the development of the logical security architecture. This information architecture should distinguish between knowledge (conceptual layer), information (logical layer) and data (physical layer).

Information itself is either static or dynamic and depending upon which of the classes it falls under, it has different needs for security to protect it.

The logical representation of the business requirements for information security is expressed through a security policy. Security policies are high-level statements of what sort of security and how much security is needed, but they do not state how that security is to be delivered.

Security policy exists at several levels of granularity and applies to many different domains. These different security policies are inter-related through a logical security policy architecture framework.

Security policy is implemented through a series of security services. These security services are logical services, specified independently of the physical mechanisms used to deliver them. They are driven from the conceptual security architecture layer above, most specifically from the Business Attributes Profile, the control objectives and the security strategies.

Security services are grouped together under headings that describe their function in the multi-tiered approach to security described in Chapter 5.

Hence there are security services for prevention, containment, detection and notification, event collection and event tracking, recovery and restoration and assurance.

Logical security architectures are created by the integration of these security services into a meaningful whole. Authentication services, directory services, certificate management services, access control services, intrusion management services and many more, all interlock to form the overall logical security architecture.

7

Physical Security Architecture

The physical security architecture is the implementer's view of life, the bricks and mortar of your enterprise security architecture. The previous chapter looked at logical functionality and flows. Now we need to look at physical boxes, how many of them, where they are located, their size and performance and how much bandwidth you need to connect them together. We also need to look at the physical data structures that are used to realize any logical information structures and at the physical security mechanisms that implement the logical security services. In this chapter you will learn about:

1. How business information at the logical layer is mapped onto data structures like databases at the physical layer
2. How the physical security mechanisms embedded within file management and database management systems can be applied to deliver the security services
3. The use of rules, practices and procedures to provide the implementation of security policies
4. The mapping of physical security mechanisms to deliver logical security services
5. How cryptographic mechanisms are used to deliver security services
6. Why vulnerabilities in security mechanisms are often difficult to foresee
7. The types of physical security mechanisms that can be used to deliver user and application security
8. The physical security mechanisms available for providing security on host platforms within the network infrastructure

7.1 Business Data Model

Please refer to Chapter 3, Figure 3.3, the SABSA Matrix, Physical layer, Asset's column, where you will see a cell entitled Business Data Model. It is not the task of a security architecture team to develop the Business

Data Model itself, this is assumed to be a pre-made model to be updated with relevant security data. However, there are many mechanisms embedded within the data storage and management systems that can be applied for security purposes and these are discussed in this section.

The logical architecture layer is concerned with information and at the beginning of Chapter 6, the differences between knowledge, information and data were discussed. Now, in the physical layer, the focus is on data.

This means that we are concerned with the physical organization and management of data so that it supports information and knowledge at the higher architecture layers. This physical data management involves:

1. File structures, including record structures and field structures
2. File management tools, including directory management
3. Databases structures
4. Database management systems

7.1.1 File and Directory Access Control

File management systems are a featured sub-system of every operating system. Typical examples are the file management capabilities within the UNIX/Linux operating system family or within the MS Windows 10/11/2022 operating system family. These types of operating systems offer discretionary access control, meaning that each file has an owner and that at the discretion of the owner the file can be shared with other users or groups of users.

Each file and each directory have permissions set to control what actions can be taken by the owner and by others who have been granted access at the discretion of the owner (Read, write, execute).

Another aspect of file security is protecting the integrity of the file by the use of a file locking mechanism, something that is often overlooked in security. This means that if one user is accessing a file, another cannot do so until the first user has closed the file. If this were not so, then different users would be making changes to the file concurrently and it would soon become corrupted as the different changes conflicted with one another.

For a wider discussion of operating system security and how these file security mechanisms fit into the bigger picture, see the Platform Security section later.

7.1.2 File Encryption

Encryption of data is discussed later in this chapter under Security Mechanisms. One of the ways that encryption can be applied is to encrypt entire files, selected records within a file or selected fields within a database record. The main challenge with using file encryption is how to manage the encryption keys. If the solution is to put the keys in another file, then no advantage has been gained. Encryption keys must be placed in a physically secure location, which implies some sort of tamper-resistant hardware device.

Alternatively, the data keys can be controlled by a master key that is derived from a passphrase1 and not stored anywhere on the system. A master key is a key-encrypting key used to encrypt a population of working keys or data keys so that they can be stored securely in an ordinary file. This functionality exists within SQL Server and Active Directory respectively. When a data key is required for use, it is retrieved from the file and decrypted using the master key. There is then only a single key, the master key, that must be given a physically secure method of storage.

If the master key is derived from a user passphrase that is not stored, there is an issue regarding the potential strength of the passphrase and the key that is derived from it. Key strength and password or passphrase strength are measured the same way – in terms of their entropy. Entropy is a measure of the randomness contained within the data. The redundancy of data reduces the entropy. Entropy is measured as the number of truly random bits in a unit of data, as opposed to redundant bits. Text that is readable by humans (such as natural-language passwords and passphrases) has high levels of redundancy in it, and hence the entropy density is low. For example, normal English-language text can be bit-encoded using ASCII to produce a string of characters, each of which has eight bits.

Various combinations of tamper-resistant hardware devices encrypted key files and passphrase derived master keys are to be found in practical key-management schemes, like the key vault in MS Azure. Cryptographic techniques can also be used to enhance file-integrity protection by computing cryptographic hashes on each file or each record and storing those with the file. If an unauthorized change is made, this can be detected.

7.1.3 Database Security

Database management systems are much more sophisticated than file systems, offering much more extensive data management and protection facilities.

One principal difference is that a database management system offers concurrent access to many users to the same data resources and must manage this without allowing the data to become corrupted by conflicting changes made by concurrent users. In a file system, the entire file is locked when a user is accessing it, but it would be unthinkable to lock the database so that only one user could make use of it. The locking mechanism must be at a much finer granularity within the database structure.

Database locking is at the record level. The hierarchy of data structures in a database is shown in

A sales database will contain several tables, including a customer table, orders table and a products table. The customer table contains a few customer records, one for each customer. The customer record comprises several fields, such as customer number, customer name, address, credit limit, etc.

The functional (logical) description of database integrity control that is achieved through the record locking mechanism is:

1. Atomicity of transactions:
2. Recoverability of transactions, the recovery sub-system of the database management
3. Serializability of transactions:

To achieve this a database management system has certain sub-systems:

1. Transaction manager
2. Scheduler – to control ordering and serialization
3. Recovery manager – to manage transaction commitment and cancellation
4. Cache manager

Database recovery is managed through a series of mechanisms:

1. Database backup to create checkpoints
2. After restoring the database to a checkpoint, providing roll-back from the saved checkpoint to a previous business position
3. After restoring the database to a checkpoint, providing roll-forward from the saved checkpoint to a future business position

7.1.4 Security Mechanisms in SQL Databases

SQL (structured query language) is the ANSI standard language that allows you to access a database. It supports:

1. Execution of queries
2. Retrieval of data
3. Insertion of new records
4. Deletion of records
5. Updating records

You can use many of the functions within SQL databases specifically to implement security mechanisms. These are summarized here:

Each user has a user account just as in an operating

1. There are three classes of users:

    1.1 System administrator – has access to and controls all databases in the entire DBMS

    1.2 Database administrator – has access to and controls a given database

    1.3 User – has access privileges as defined by the system administrator or the database administrator.

2. Groups are used with group identifiers and group passwords
3. A user is often mapped to an application that makes automated database access
4. User and group privileges granted by an administrator can be assigned to:

    4.1 Specific tables

    4.2 Specific views

    4.3 Specific procedures.

5. In an SQL program the commands for managing user privileges are GRANT and REVOKE
6. SQL supports mechanisms to protect the integrity of data in the database
7. Creating views with the WITH CHECK option to force underlying table updates to conform to constraints on value ranges
8. COMMIT and ROLLBACK functions to allow mistakes to be corrected

You can use views to provide restricted access to a table. This provides a mechanism for fine control over access to users. The user gets access only to the view, which is a sub-set of data in the main table. It might contain only certain records or only certain fields.

You can define as many views as needed for data access and these may overlap one another in any way, so there is great flexibility to provide access rights to individual users or groups of users.

Stored procedures are sets of SQL commands bound together and executed as a whole. This allows you to define whole tasks and make them available to users without giving them the use of the individual raw commands. Thus, you can grant quite sensitive operations to a user but only in the context of the stored procedure and thus under close control. There are many more tools in the database toolbox to control access to information and queries against the database. There are whole books dedicated to this for all of the major database systems out there, MS SQL Server, Oracle, MySQL, ...

7.1.5 Distributed Databases

A distributed database is one that is logically a single database but is physically distributed over several servers, perhaps at several geographically separated sites if located within a cloud datacenter by one of the big cloud vendors. This brings some additional challenges for managing database integrity and consistency.

The main problem is that one part of the distributed system can fail independently of any other part. So, to maintain the concept of atomicity and consistency of transactions we need some additional functionality, since if one part of the database has committed a transaction and another part has not, then they will become desynchronized and inconsistent.

7.2 Security Rules, Practices and Procedures

Please refer to Chapter 3, Figure 3.3 the SABSA Matrix, Physical layer, Motivation column, where you will see a cell entitled Security Rules, Practices and Procedures. This section looks at aspects of this deliverable. At the logical security architecture layer, you have security policies, as discussed in Chapter 6. These now get turned into sets of rules and into practices and procedures at the physical architecture layer.

7.2.1 Security Rules

A rule is a specific filter against which automated decisions are made by security sub-systems. For example, rules are used in the following types of security sub-system:

1. Firewall rules
2. Database rules
3. File system rules

Rules are often built into access control lists. An access control list is made up of access control entries, each of which contains one or more rules (read, write, execute). The key difference between security rules and security policies is that whilst policies may require interpretation, rules do not, they are concrete and enforced.

7.2.2 Security Practices and Procedures

Security practices are a generic description of objectives in security management. Usually, the term practice is used as part of a phrase such as 'good practice' or 'best practice'. This implies adopting an approach that has been found to work well in similar circumstances.

Security practices are more specific than security policies and imply a definite behavioral content to them, but they are not as specific as security procedures. Security procedures are documented step-by-step instructions on how a specific security task is to be performed. A procedure is specific to a particular platform or product and deals with the minutiae of how that specific technical environment works and is maintained. A new release of a given product may require the procedure to be updated because the product is now slightly different from the previous release and needs a different set of steps. In the procedure.

So, a procedure for backing up data on one type of platform will not be applicable to another different platform and would need to be modified to be useful. However, the practices of data backup will be the same for both computers because they involve the same generic steps.

Another term that is used in this context is guidelines. These are generally non-specific, but they tend to be more in the style of giving good advice to people about how to behave to comply with policy and to exhibit good practice. Guidelines require intelligent interpretation and some training will be required in order for the interpretation to be uniform.

7.3 Security Mechanisms

Please refer to Chapter 3, the SABSA Matrix, Physical layer, Process column, where you will see a cell entitled Security Mechanisms. In this section, the most common security mechanisms are mapped to the logical security services already introduced in Chapter 6. Some of these security mechanisms, mostly those with a cryptographic basis, are discussed in greater detail in this section. Others, such as data management mechanisms, are discussed in other parts of the chapter.

A security mechanism is a physical means by which a logical security service is implemented. Some security mechanisms were already been discussed in the earlier sections of this chapter, like databases and filesystems. Some security mechanisms have been classified in ISO standards. However, there are many more standards and mechanisms, with more arriving every year, that you will encounter that will also serve us well.

7.3.1 Mapping Security Mechanisms to Security Services

Table 7.1 shows the mapping of a subset the security mechanisms described here to the security services that are described in Chapter 6.

Many of the security mechanisms listed in Table 7.1 are simple and obvious in their nature, and there is no reason to explain each one in detail. Other mechanisms.

Table 7.1 Mechanism/services mapping

Logical security services physical security mechanisms

Logical Security Services	Physical Security Services
Entity Unique Naming	Naming standards
	Naming procedure
	Directory system
Entity Registration	Registration policy
	Registration authority system
	Registration procedure
Public Key Certification	Certification policy
	Certification authority system
	Certification procedure
	Certificate syntax standards
	Certificate publishing mechanism (directory)
	Certificate revocation list (CRL)
	CRL publishing and management (directory)
Credentials Certification	See Above
Directory Service	Directory system
	Directory access protocols
	Directory object and attribute syntax rules
	Directory replication
Entity Authentication	Login procedure
	User passwords and tokens
	Client user agents for authentication
	Authentication exchange protocols
	Authentication server system
	Directory system
Session Authentication	Mutual two-way and three-way authentication exchanges
	Session context
Message Origin Authentication	Message source identifiers, protected by:
	Message integrity checksums
	Digital signatures
	Hashing
Message Integrity Protection	Message integrity checksums
	Digital signatures
	Hashing
Message Replay Protection	Message nonce values protected by message integrity
	Checksums
Message Contents Confidentiality	Message contents encryption
	Encryption key management
	Routing control to physically secure networks
Non-Repudiation	Digital signatures
	Notarization servers
	Transaction logs
	Trusted third party certification and arbitration
Traffic Flow Confidentiality	Traffic padding
Authorization	Roles
	Fixed role associations with entities
	Real-time role association with entities
	Authorization certificates
Logical Access Control	Local access control agents
	Local role access control lists (ACLs)
	Central access manager (CAM)
	CAM role ACLs
	Central application access control agents
	Central application role ACLs
	Database management system mechanisms
	File system mechanisms

Table 7.1 Continued

Audit Trails	Event logs
	Event log integrity protection mechanisms
	Event log browsing tools
	Event log analysis tools
	Reporting tools
Data at Rest	Logical access control mechanisms
	Physical access control mechanisms
	Stored data encryption
	Media storage security
	Media disposal procedures
Stored Data Integrity Protection	Message integrity checksums
	Digital signatures
	Hashing
Software Integrity Protection	Development lifecycle controls
	Delivery and installation controls
	Production system configuration control
	Production systems change control
	Production system management authorization
	Crypto checksums on object code images
	Regular inspection of object code images and checksums
	Anti-virus tools
System Configuration Protection	Production system configuration control
	Production systems change control
	Production system management authorization
	Cryptographic checksums on configuration data files
	Regular inspection of configuration data files and checksums
Data Replication and Backup	Regular backup copying
	Backup media management: labelling, indexing, transport, storage, retrieval, media recycling, media disposal
Trusted Time	Secure time server with clock
	Secure time server protocols
Security Policy Management	Data content monitoring and filtering
	Real-time system monitoring
Security Service Management	Security service management sub-system
	Secure management protocols
	Management agents in managed components
	Access control at all agents and sub-systems security alarms
Security Training and Awareness	Training courses
	Training manuals and documentation
	Publicity campaigns
Security Operations Management	Operator authentication mechanisms
	Operator activity logs
	Operations event logs
Security Provisioning	Security service management sub-system
	Secure management protocols
	Management agents in managed components
	Access control at all agents and sub-systems security alarms
Security Administration	See Above
Security Monitoring	User activity logs
	Application event logs
	Operator activity logs
	Management event logs
	Event log browsing and analysis
	Reporting
	Real-time system monitoring and alarms
Security Measurements and Metrics	Cryptographic test mechanisms
	Inspection tools
	Penetration testing
	Statistical tests
Intrusion Detection/Prevention	Intrusion signature analysis on network traffic
	Real-time system monitoring
	Alarms

Many of the security mechanisms listed in Table 7.1 are simple and obvious in their nature, and there is no reason to explain each one in detail. Other mechanisms.

However, some of the security mechanisms are technically detailed in nature, especially those of a cryptographic or intrusion prevention nature. The following sections describe a few selected mechanisms in slightly greater detail. For more detailed information on other security mechanisms, and indeed for full details of those mechanisms described below, you are recommended to research in other books more focused at this level of detail and with vendor organizations who provide tools and products that contain these mechanisms. Some of them will be detailed and described in Chapter 13, where is use examples from Cisco and Microsoft.

7.3.2 Cryptographic Mechanisms and Their Uses

Cryptography has some very specific roles to play in securing information. There are four fundamental security services that can be implemented using cryptography:

1. Confidentiality, preventing the unauthorized disclosure of information
2. Integrity, protecting information content from being altered in any way without this being detected
3. Authenticity, proving that information originated from a trusted source
4. Non-repudiation, preventing a dishonest party from later denying the authenticity of information.

There are various ways to present these services, as we can see in Table 7.1 above. However, if the security service you want to provide cannot in some way fit into the four categories above, then cryptography is not the right mechanism for your solution.

The strength of cryptographic mechanisms is mostly related to the number of bits in the cryptographic key and hence the number of possible key values. This assertion assumes that there are no specific weaknesses in the algorithm and that the cryptanalyst must rely on searching the entire key space (brute force) to find the key value being used.

The main constraint on using cryptographic mechanisms in information systems is the need to manage the keys. This must be done securely and efficiently. This is discussed in a little more detail in a later section below.

7.3.3 Encryption Mechanisms

An encryption mechanism transforms original raw data into an encrypted form of the data. If a good quality encryption algorithm is used the transformation is complex and opaque, such that it is infeasible for an opponent to analyze the ciphertext to discover the original plaintext. In most cases, the transformation is controlled by an encryption key. This is an additional piece of data that influences the transformation. For a given plaintext if you change the key you will obtain a different ciphertext.

If the purpose of using the encryption mechanism is to hide the plaintext from an opponent, during transmission over a network or perhaps during storage on a computer system, but later to recover the plaintext in its original form, then the encryption must be reversible. The reverse process is called decryption. In some cryptosystems, the key used for decryption is the same as the one used for encryption and is known as symmetric cryptosystems.

However, there is a class of cryptosystem where the reverse transformation requires a different, but mathematically related key and these are called asymmetric for that very reason. The asymmetric algorithms are also commonly known as public key cryptosystems because the encryption key can be published and only the decryption key needs to be kept private.

Not all cryptosystems are reversible and there are good applications for one-way functions as cryptographic algorithms. The storage of passwords in a password file is one such example. You want to make sure that if someone steals the file, they cannot reverse the encryption and recover the plaintext passwords. The way to check a submitted password is to encrypt it and compare the ciphertext result with the stored ciphertext in the password file. Such a system is still vulnerable to so-called dictionary attacks, but direct decryption of the stored password is not feasible.

7.3.4 Data Integrity Mechanisms

Cryptography can be applied to generate secure checksums or seals that provide a mechanism for detecting modifications to the original data. These seals are also known as message authentication codes (MACs) or message integrity checksums (MICs) in some contexts. The method used is to process the entire data contents to be protected through a cryptographic transformation that produces a short checksum or hash value.

If the checksum algorithm does not require a cryptographic key, then the method is called hashing. Hash values are usually used within digital signature functions to improve overall efficiency and speed, but there is also a variation of the MAC technique called HMAC, also known as keyed hashing.

As with any application of cryptography, the method works only if a key management service is provided to make the keys available wherever they are needed while keeping them a secret from an opponent.

7.3.5 Public Key Certificates

Where asymmetric cryptographic techniques are used each participating entity has a key pair. One of these keys of the pair is published and available to all other participants. This key is known as the public key. The other key of the pair must be kept completely private and not disclosed to any other participant, called the private key.

Although public keys are known to all and therefore not confidential, they must be authentic. An opponent could steal your identity and publish his own public key pretending to be you unless there is some mechanism to prevent this from happening. The mechanism used is called certification.

Every public key in the community must be certified by a trusted authority. The trusted certification authority digitally signs each public key to create a public key certificate. It is the certificate that is published for all to use. To be sure that you are using an authentic public key, you must first check the signature of the certification authority in the certificate, to make sure that it is a CA that you trust.

7.3.6 Digital Signature Mechanisms

Digital signatures are based on the application of asymmetric cryptographic techniques. The private key is used to create the signature and the public key is used to verify the signature. Public keys must be certified as described above. Digital signatures prove both the integrity of the data content and the authenticity of the data source. The signatures are also non-repudiable.

To apply an asymmetric cryptographic algorithm directly onto the target data would in most cases be computationally inefficient, take a long time to execute and create a signature the same size as the original data. To create an efficient practical digital signature system, you use a two-stage process. First, the data is hashed to create a short hash value. This hash value is digitally signed and attached to the message by the signatory. The verifier must rehash the message and recover the original hash from the received signature, then compare these two hash values. If they are the same, there is a high level of assurance that the signature was made by the authentic signatory.

7.3.7 Cryptographic Key Management Mechanisms

Throughout the previous sections, I have pointed to the need for efficient, secure key management. One could write an entire book of similar size to this one just on this topic alone and books on the subject are out there. There are however a few things that should be said in a book of this type. There are some general principles of key management that are worth considering:

1. However strong the cryptographic algorithms might be, if the keys are poorly managed and can be learned by an opponent, the overall system is weak.
2. In general, it is a bad idea to allow human beings to have knowledge of cryptographic keys. Humans are often the weakest link in any security system.
3. It is often impossible to avoid some human knowledge of some keys somewhere in the system, usually to ensure that when all the technology has broken you can still recover the system manually.
4. Key management should in all other ways be completely automated, using key encrypting keys to move working data keys around from one place to another.
5. A truly random number generator should be used to generate new keys to ensure that they are not predictable.
6. Tamper-resistant devices can protect keys from unauthorized disclosure.
7. Keys should be limited in their exposure to cryptanalysis. This exposure is defined in terms of both the number of ciphertext samples that an opponent can collect and the time that the opponent must work on those samples.
8. Different keys should be used for different functions, different parties and different channels to ensure cryptographic separation of logical communications.

Key management covers the entire lifecycle of keys and everything that happens from them from birth to death:
Generation of keys.

1. Communication and distribution of keys
2. Storage of keys
3. Entry and installation of keys
4. Checking the validity of keys
5. Usage of keys
6. Changing the active key

   • Archiving of keys

7. Destruction of keys
8. Audit of key operations and usage
9. Key backup and recovery

7.3.8 Cryptographic Services Physical Architecture

There are two fundamentally different physical architectures for providing cryptographic services in a distributed system:

1. The in-line architecture, providing cryptographic services in the networking layers of the system architecture
2. The on-line architecture, providing cryptographic services at the applications layer of the system architecture

The in-line cryptographic architecture can provide the following types of network security service:

1. Physical layer confidentiality is implemented by encrypting the entire bitstream using in-line pairs of link encryption units.
2. Link-layer confidentiality is implemented by encrypting the payloads in Ethernet packets, using functionality embedded in the network interface card.
3. Network layer confidentiality and integrity checking are implemented by encryption and integrity checksums over the payloads in packets of the network protocol. IPsec transport mode services are of this type
4. Transport layer session authentication implemented with cryptographic authentication exchanges. Additionally, transport packet payload confidentiality can be implemented by encryption. TLS services are of this type

Key management for the in-line cryptographic architecture is most likely to be on an in-line peer to-peer basis, although for OSI layer 3 encryption systems there may be a key distribution server. The on-line architecture can support these application-level services:

1. Confidentiality of transmitted data in the payload of application data structures. The confidentiality can apply to the entire payload or to selected fields within it.
2. Integrity checking of transmitted data in the payload of application data structures.
3. Authentication of the source of transmitted application data structures.
4. Peer-to-peer application-level authentication using cryptographic authentication exchanges.
5. Non-repudiation of transmitted application data structures.
6. Encryption of application data stored in files or databases.
7. Integrity checking of application data stored in files or databases.
8. User and other entity authentication services based on cryptographic authentication exchanges,

The key management for the cryptographic architecture can be and likely is, highly sophisticated, including the storage of large numbers of keys in the data store. To keep the stored keys secure they are encrypted under a higher-level key-encrypting key, often named a master key. The master keys can be stored in a physically secure, tamper-resistant hardware unit, which may also offer accelerated processing for cryptographic computations and a hardware random number generator for key generation. This type of special hardware unit is sometimes known as a hardware security module (HSM). There are many refinements that can be incorporated into this generic on-line architecture. It offers by far the most flexible approach to providing cryptographic services to applications, although at a significant cost. Typical applications where these type of on-line cryptographic architecture are used are:

1. E-mail and all similar store-and-forward messaging applications
2. Interactive messaging services
3. File transfer applications
4. Financial payment systems and money transfer systems in banking applications
5. Point-of-sale and cash machine authorization systems
6. Secure eCommerce, eProcurement and similar applications.

7.3.9 Other Cryptographic Mechanisms

The discussion here has mentioned a few of the most encountered cryptographic mechanisms, but there are many more that you may encounter. Some of these are listed here:

1. Zero knowledge proofs

   - Secret sharing schemes

2. Multi-party signatures
3. Dual signatures
4. Blind signatures
5. Elliptic curve cryptography

7.4 User and Application Security

Please refer to Chapter 3, Figure 3.3 the SABSA Matrix, Physical layer, People column, where you will see a cell entitled Users, Applications and the User Interface. This section describes the security mechanisms that are commonly applied to implement user and application security. Among these mechanisms is the user password, which is an important element of the security user interface and there is some discussion of the issues that surround the use of this mechanism.

In Chapter 6 the security services applicable to users and applications are discussed at length. The security mechanisms by which these services are implemented are straightforward. They are:

1. Directory Mechanisms

   1.1 Object class definitions
   1.2 Attribute syntax definitions
   1.3 Directory access protocols
   1.4 Directory access control mechanisms
   1.5 Directory user authentication mechanisms
   1.6 Inheritance checking mechanisms.

2. Central Access Manager Mechanisms
3. Database Mechanisms
4. File System Mechanisms
5. Operating System Mechanisms
6. Application Mechanisms
7. User Authentication Mechanisms.

7.5 Platform and Network Infrastructure Security

Please refer to Chapter 3, Figure 3.3, the SABSA Matrix, Physical layer, Location column, where you will see a cell entitled Platform and Network Infrastructure. This section describes the most important security mechanisms that are used to provide security within the platform and network infrastructure. It also discusses aspects of the physical layout and topology of this infrastructure.

7.5.1 Resilience

Physical network and platform infrastructure should be built in resilient configurations to incorporate a degree of fault tolerance. The amount of fault tolerance (and hence the cost of providing it) will depend upon the business requirements for resilience and continuity of operations. The main principles of resilient design are:

1. Avoidance of single points of failure by ensuring that there is always an alternative mechanism for delivering a given function or service.

2. Redundancy of hard physical components such that if one fails another is available
3. Backup and restoration procedures for all soft physical components
4. Recovery procedures worked out in advance for foreseeable failure scenarios
5. Automated recovery and reconfiguration where possible
6. Extensive event logging, monitoring and reporting to help foresee possible failures before.

In applying these principles to the design of network topologies, a few specific approaches are commonly used:

1. Multiple communications cables and channels, often with diverse physical routing
2. Separation of cable routes from buildings to avoid all the cables suffering physical failure
3. Alternative telco carriers where local market conditions allow this
4. Dynamic automated re-routing and re-configuration to create a self-healing network
5. Regular testing and monitoring of these various resiliency features to ensure they are operating correctly
6. Physical and environmental security of communications rooms and computer rooms.

The same principles are also applied to providing highly resilient host platform facilities for applications:

1. Dual processing facilities in separate data centers, often geographically separated by several hundred kilometers
2. Fault-tolerant computer systems or middleware that automatically organizes data mirroring or distributed processing.

7.5.2 Performance and Capacity Planning

The whole issue of system performance has to do with being able to deliver both the processing power and the communications bandwidth required to handle the volumes of information to be processed and transported. So, the entire disciplines of capacity planning and performance tuning are relevant to the overall provision of secure, resilient information-processing services to meet the business requirements. This has become especially important with the inclusion of cloud services into the infrastructure.

7.5.3 Platform Security

Platforms are the host computers. The host can be anything from a small personal computer up to a super-computer. The word host does not imply any specific size or type of computer; merely that it can host some application or virtualization software.

A platform is generally thought of as a combination of the underlying hardware together with the operating system. Hence people talk about a Linux platform or a Windows server platform. Platform security is therefore focused upon how the operating system should be configured and operated to meet the business requirements for security. Here some of the main functions and mechanisms available for managing platform security are summarized.

The generic file level-security mechanisms and process-security mechanisms that we find in all commercial operating systems are based on a discretionary access control (DAC) policy and include:

1. Each real user associated with a user account to control access by that user
2. Unique username and identifier for each user account
3. Groups of users with a group name and group identifiers
4. Password for each user account or a group account
5. A super-user account that has access to everything and controls the system
6. Mapping of physical devices ontological file structures to control access to these devices
7. Files and directories created by a user being owned by that user who can grant access privileges to other users
8. Differentiated access privileges
9. A home directory for each user, increasingly a cloud option like OneDrive
10. An ability for some programs to give the user a higher privilege value than that of the person who runs the program to grant higher privileged access under the careful control of an approved program
11. Event logs at various levels capture event data for investigations and forensic analysis

Some of the major security issues in these types of operating systems are as follows:

1. The super-user is all-powerful and can do anything to any file in any directory.

2. Carelessly coded programs that allow the user identifier to be set to a higher privilege level than that of the process owner are one of the ways in which an opponent can gain higher privileges.
3. Trojan horse attacks are a major source of threat. If an opponent can arrange for a Trojan version of a highly privileged system utility to be run (possibly with super-user privilege), then the opponent can take control of the system.
4. Writeable directories are a major vulnerability since they allow files to be added or removed from the directory.

To run a secure production environment for business applications, these are some of the good practice guidelines for managing the platform:

1. There should be a written security policy supported by detailed standards aimed at those with operational responsibility for maintaining the operational security of the platform.
2. File and directory permissions should be carefully designed as part of the application to meet the business requirements.
3. There should be stringent change control with all executable files.
4. There should be regular scanning to detect changes in any authorized executables and to detect the unauthorized installation of any other executables.
5. There should be strict control over maintenance procedures
6. All non-production executable software should be removed from the production platform.
7. There should be a strict process for accepting and releasing new versions of production software.
8. All default system accounts should either be removed completely or at the very least have their passwords changed
9. The super-user account *must not* be used for routine operations.

7.5.4 Hardware Security

A major problem that has dogged computer security since the first computers were developed has been how to provide security at the physical hardware level. Something that has become a bigger issue in recent years, with the finding of security weaknesses within CPU's. Whatever logical control systems have been built in software, if you cannot control the hardware, you cannot control the software execution environment and hence it is always possible for unauthorized software to be loaded and executed, taking overall control of the machine and negating much of the security provided through other authorized software. The successful exploitation by computer viruses, Trojan horses and hacking is a direct result of this problem. There are two ways in which a certain amount of hardware security has been achieved in the past:

1. By physically surrounding the entire computing platform and its peripherals in a secure building
2. By building tamper-resistant computing equipment, in which the physical boxes of the computers are protected from unauthorized use.

Whilst these methods work, they have severe limitations:

1. The use of a tamper-resistant module (TRM) for providing the traditional host security module (HSM) architecture still requires the application software and the main host platform to be surrounded by a physically secure computer suite, otherwise, the application software can be compromised.
2. The secure computing suite is inappropriate when the platform to be secured is a personal computing device.

The way to solve these problems in the personal computing environment is to develop new hardware architectures that incorporate inexpensive physical security at the chip level.

7.6 To Summarize: Physical Security Architecture

Business data is managed by means of file management, database management and directory management systems, all of which have many mechanisms built into them that can be applied for implementing security. Physical data storage systems can be designed to ensure that data is not lost during a physical failure of an individual data storage device.

Security rules, practices and procedures, integrated with the application of various security mechanisms, are used as the physical implementation of logical security policies.

Security mechanisms in general can be mapped onto the set of logical security services developed at the logical security architecture layer.

It is almost impossible to create a security mechanism that does not have some form of vulnerability and often these vulnerabilities are difficult to predict until an opponent finds a means to exploit them. Cryptographic mechanisms are applied to provide high levels of assurance of confidentiality, authenticity, integrity and non-repudiation. These mechanisms cover encryption and authentication, using both symmetric and asymmetric cryptosystems. They also require cryptographic key management mechanisms and physical hardware security mechanisms to protect cryptographic keys in storage and in use.

Security mechanisms for protecting user privileges and the applications that they use are mainly focused on user authentication and access control. The user password is an important and ubiquitous security mechanism, but it has many issues that need to be considered carefully. Tokens and biometrics can be used to strengthen user authentication procedures.

The security mechanisms applied in the ICT infrastructure include those deployed to manage resilience, capacity and performance. They also include the entire range of possibilities for physical layout design and topology. The embedded security mechanisms within operating systems are being used to control access to platforms of all types, but they tend to be impotent against the threat posed by malicious software.

Physical protection from interference with software running on a computing platform (such as the unauthorized installation of malicious software) is achieved through physical security of sites, buildings and rooms, physical access control for authorized personnel, robust procedures for software installation and release and tamper-resistant equipment for specialized applications.

In the future it is likely that there will be a significant shift towards the use of tamper resistance at the chip level, allowing physically secure hardware platforms to be deployed for personal use, including PCs, Tablets and mobile phones. This has brought about a major change in the ability to defend against malicious software threats.

8

Component Security Architecture

The component security architecture is the tradesman's view of life – the specialized tools and product components of your enterprise security architecture. This chapter looks at a selection of these components but stops short of discussing any specific brand or vendor. This is because:

1. It is not appropriate to comment on one versus another here and
2. The marketplace changes so rapidly that the authors have focused on writing as timelessly as possible by sticking to generic principles.

To find specific tools and products you must do your own research. As before, the chapter structure follows the cells in the Security Architecture Matrix at this layer, although at this level there is little to say on most of the cells. In this chapter you will learn about:

1. How standards are needed to achieve consistency and inter-operability between security architecture components
2. The role of ASN.1 and XML as fundamental syntax standards on which many other standards are built;
3. The major international, national and industry sector standards-making bodies and their contributions in providing security-related standards
4. Functional security standards based upon XML, including web services comprising various modular building blocks and protocols.

8.1 Detailed Data Structures

Please refer to Chapter 3, Figure 3.3, the SABSA Matrix, Component layer, Asset's column, where you will see a cell entitled Detailed Data Structures. This section discusses the basic syntax standards that are used to create standardized data structures for the security-related protocols that are used to exchange this data.

161

8.1.1 Inter-Operability

At the component layer of any architecture, it is essential that the individual components selected from different vendors should be capable of being plugged together to build integrated structures, something that has historically been difficult to achieve. This means that they must have compatible interfaces otherwise, systems integration becomes very difficult.

The best way to achieve integration of individual components is to select components that have compatible interfaces, usually internationally standardized interfaces or de facto industry standardized interfaces to support ease of integration and inter-operability. This quality should be one of several that guide your search for suitable tools and products.

One specific area of standardization is in the data structures that your components exchange. If they do not agree precisely on syntax rules and protocols, then they will not communicate successfully. This is where you need to look for compliance with recognized standards, whether they are formal.

8.1.2 ASN.1

One important international standard is abstract syntax notation number one (ASN.1). ASN.1 is described in ISO/IEC 8824. It is a strongly typed language specifically designed for specifying application-layer protocol data units for communications protocols of all sorts.

The language has some built-in types that act as the building blocks. These include the simple types: BIT STRING, BOOLEAN, CHARACTER STRING, ENUMMERATED, INTEGER, OCTET STRING, NULL and REAL. If you are familiar with programming, then these should be known to you.

Other built-in types include UTC Time, Co-ordinated Universal Time (GMT) and Generalized Time, which is a local date and time including a difference from GMT.

The CHARACTER STRING type also has various in-built classes defined, including Graphic String, IA5String2, Numeric String, Printable String, Teletex String, Videotex String and Visible String.

The OBJECT IDENTIFIER provides an external reference to an ASN.1 object that has been described and registered by one of several international authorities. The reference itself appears as a series of integer numbers, which are the node numbers on a hierarchical tree structure beginning at the root. So, the OBJECT IDENTIFIER for the MD54 algorithm is {1,2,840,113549,2,5}. Figure 13.1 shows how this string is constructed from the tree.

Also included in the ASN.1 standard are the encoding rules, which can be the Basic Encoding Rules (BER) or the Distinguished Encoding Rules (DER). The DER definition represents a later improvement (in 1993) that guarantees a unique encoding of every object. Actual encoding is achieved with special compilers. ASN.1 can come across as complex and, in all fairness, it can be! But you will not need a detailed understanding of ASN.1 in your Security Architect career.

8.1.3 Extensible Markup Language (XML)

XML is a meta-language that allows you to create specialized application-level languages for specialized client-server interactions. 'Documents' are created using these specialized languages and there are many different document types out there for many different purposes.

In this context, document refers to any structured information and so can describe any information that needs to be exchanged for Business purposes. Thus, in addition to text, a document may contain graphics, mathematical equations, structured numerical information such as tables and charts and much more.

In effect, XML has become the new generation mechanism for what was originally called electronic data interchange (EDI). However, XML is many times more powerful than the old EDI syntax because of its ultimate flexibility, even though it is slowly being exchanged for JavaScript Object Notation (JSON). Also, whereas originally EDI was focused on simple point to-point file transfers, XML is supported by the more powerful web technologies, you can even store XML and JSON documents in databases now.

So, where those activities require protection of confidentiality, integrity, authenticity, nonrepudiation, authorization, accountability and many other similar attributes, mechanisms are needed within XML to deliver that protection.

There are already many such specific mechanisms being developed within the XML standards, beginning with simple building blocks such as XML Encryption and XML Signature. These and others are described in more detail in later sections of this chapter.

8.1.4 Relationship between ASN.1 and XML

ASN.1 and XML are neither in conflict nor in competition. They do different jobs in different ways. ASN.1 was developed as a tool for specifying data communications protocols in a traditional sense, whereas XML is intended for use in encoding document exchanges. The difference is subtle because an XML document exchange is in fact a very elaborate data communications protocol.

However, there is another very clear difference, in that ASN.1 produces binary encoded data, whereas XML produces text-encoded data.

However, the bandwidth revolution in which huge bandwidths have become available at low cost has rather taken away the anxiety about efficiency. Where efficiency is more of an issue because the communications protocol is implemented in a realtime

embedded system such as a router, ASN.1 has an important role. Where the interaction is truly in the business applications layer, XML provides much greater flexibility, the bandwidth is available to support it and performance is not so critical.

Additionally, because XML can embed any type of data, including binary data objects, you will also see examples of ASN.1 data structures inside XML documents. A particularly relevant example is the X.509 digital certificate, which is specified in ASN.1 and binary-encoded but which will often be sent as part of an XML document.

8.1.5 Standard Security Data Structures

There are many security-related data structures that are already or will be in the future standardized. These types of structure include:

1. Digital signatures
2. Digital certificates
3. Certificate management protocols
4. Time protocols
5. Authentication exchanges

The possibility now exists to describe large complex documents in standardized formats that can be created and interpreted by automated processes.

8.2 Security Standards

Please refer to Chapter 3, Figure 3.3, the SABSA Matrix, Component layer, Motivation column, where you will see a cell entitled Security Standards. This section summarizes the main standards-making bodies that operate in information security and provides an overview of the focus of each body.

If every component were built to a unique interface specification, nothing would ever work together. So, standards are needed to ensure that many different components can be integrated to form a larger system. Having agreed that a standard should make everything the same, you then find that there are so many standards from which you can choose, that is almost feels like a jungle with no standardization!

There are literally hundreds and hundreds (possibly thousands by the time you read this) of internationally recognized standards for various aspects of security. In this chapter, the various standards-making bodies and types of standards that they address5 are summarized. There is no attempt to include a comprehensive list of all the standards because (1) such lists are not very interesting and (2) they go out of date very quickly as new standards arc published. To find out the most recent standards you are advised to look up the current activities under each of the standards bodies listed.

8.2.1 International Organization for Standards (ISO)

The ISO is a worldwide federation of national standards bodies from more than 100 countries, one from each country. ISO is a non-governmental organization established in 1947. The mission of ISO is to promote the development of standardization and related activities in the world with a view to facilitating the international exchange of goods and services and to developing cooperation in the spheres of intellectual, scientific, technological and economic activity.

ISO's work results in international agreements that are published as international standards.

There are many ISO standards that address information security and the security of information systems and networks, some of which are referred to at various points throughout this book. ISO is behind the ISO 27001 standard for security management.

8.2.2 International Electrotechnical Commission (IEC)

The IEC is the leading global organization that prepares and publishes international standards for all electrical, electronic and related technologies. These serve as a basis for national standardization and as references when drafting international tenders and contracts.

Through its members, the IEC promotes international co-operation on all questions of electrotechnical standardization and related matters, such as the assessment of conformity to standards, in the fields of electricity, electronics and related technologies.

The IEC charter embraces all electro technologies including electronics, magnetics and electromagnetics, electroacoustic, multimedia, telecommunication and energy production and distribution, as well as associated general disciplines such as terminology and symbols, electromagnetic compatibility, measurement and performance, dependability, design and development, safety and the environment.

Where the activities of the IEC and ISO overlap, standards are published jointly. So, many of the standards around information and communications technology are joint ISO/IEC standards and this includes the security-related standards. IEC is behind IEC 62443, a security standard for SCADA/ICS systems. As mentioned elsewhere, the security of these systems as they get more and more integrated with IT systems is becoming a critical social issue! If you are working within SCADA/ICS, then this standard is well worth a read.

8.2.3 Internet Engineering Task Force (IETF)

The IETF is a large open international community of network designers, operators, vendors and researchers concerned with the evolution of Internet architecture and the smooth operation of the Internet. It is open to any interested individual. The actual technical work of the IETF is done in its working groups, which are organized by topic into several areas (e.g., routing, transport, security). Much of the work is handled via mailing lists. The IETF holds meetings three times per year.

The IETF working groups are grouped into areas and managed by area directors (ADs).

The Internet Architecture Board (IAB) provides architectural oversight. The IAB also adjudicates appeals when someone complains that the IESG has failed. The IAB and the IESG are both chartered by the Internet Society (ISOC) for these purposes. The general area director also serves as the chair of the IESG and of the IETF and is an ex-officio member of the IAB.

Each distinct version of an Internet standards-related specification is published as part of the Request for Comments (RFC) document series. This archival series is the official publication channel for Internet standards documents and other publications of the IESG, IAB and Internet community. The current index of RFCs and downloadable copies of individual RFCs are to be found on the IETF website, together with details of ongoing work programs.

8.2.4 American National Standards Institute (ANSI)

The ANSI is a private, non-profit organization that administers and coordinates the US voluntary standardization and conformity assessment system. The institute's mission is to enhance both the global competitiveness of US business and the US quality of life by promoting and facilitating voluntary consensus standards and conformity assessment systems and safeguarding their integrity. ANSI has been active in standardizing many areas of information security.

8.2.5 International Telecommunication Union (ITU)

The ITU is an inter-governmental organization established in 1865 when international telegraphy was new. When, in 1947, the United Nations was formed, the ITU became a specialized agency within the UN responsible for telecommunications.

Thus, it represents the public telecommunications interests of more than 170 countries. Amongst its activities is the establishment of standards.

8.2.6 Institute of Electrical and Electronics Engineers (IEEE)

IEEE is a non-profit, technical professional association of more than 377,000 individual members in more than 150 countries. Through its members, the IEEE is a leading authority in technical areas ranging from computer engineering, biomedical technology and telecommunications, to electric power, aerospace and consumer electronics, among others. The IEEE has nearly 900 active standards with 700 under development.

In the area of local area networking, many IEEE standards have been adopted by ISO/IEC. Ethernet standards are developed and maintained by IEEE.

8.2.7 Information Systems Audit and Control Association (ISACA)

With more than 140,000 members in over 100 countries, the Information Systems Audit and Control Association§(ISACA§) is a recognized global leader in IT governance, control, and assurance.

Founded in 1969, ISACA sponsors international conferences, training events and a global knowledge network (K-NET), administers the globally respected Certified Information Systems Auditor^{TM} (CISA§) designation and the Certified Information Security Manager^{TM} (CISM^{TM}) designation and develops globally applicable information systems (IS) auditing and control standards. Full disclosure. I am a member of ISACA and I possess the previously mentioned certifications, including a few more from ISACA.

8.2.8 Object Management Group (OMG)

Founded in April 1989 by eleven companies, the Object Management Group^{TM} (OMG^{TM}) began independent operations as a not-for-profit corporation. Through the OMG's commitment to developing technically excellent, commercially viable and vendor independent specifications for the software industry. The OMG is moving forward in establishing the Model Driven Architecture^{TM} as the 'Architecture of Choice for a Connected World'^{TM} through its worldwide standard specifications including CORBA§, CORBA/ IIOP^{TM}, the UML^{TM}, XMI^{TM}, MOF^{TM}, Object Services, Internet Facilities and Domain Interface specifications.

8.2.9 The World Wide Web Consortium (W3C)

The W3C develops interoperable technologies (specifications, guidelines, software and tools) to lead the web to its full potential. W3C is a forum for information, commerce, communication and collective understanding. It was created in October 1994 to lead the World Wide Web to its full potential by developing common protocols that promote its evolution and ensure its interoperability. W3C has around 450 member organizations from all over the world.

There are many more standards out there, some of them from the various vendors themselves, others will be internal to the organization. Which ones to adhere to will depend on the situation and the regulatory and compliance needs of the organization.

8.3 Security Products and Tools

Please refer to Chapter 3, Figure 3.3, the SABSA Matrix, Component layer, Process column, where you will see a cell entitled Security Products and Tools. This section provides a high-level taxonomy of security products and describes the common features and functions of each one.

Table 8.1 states a few of the most common types of security tools and products and gives an overview of the most found features of those components.

Table 8.1 Security tools and products

Component Type	Features/Mechanisms
Anti-piracy tools	Preventing the illegal copying and distribution of software
Anti-theft devices	Preventing the theft of equipment items such as PCs
Anti-virus scanners	Scanning for known viruses and other malicious software and repairing any damaged files
Biometric devices	Providing personal authentication based on measurement of a bodily feature
Boot-protection software	Preventing the booting of a PC from a diskette to get unauthorized access to the hard drive
Business continuity planning and disaster recovery planning tools	Supporting the collection and management of planning information
CCTV monitoring	Physical site surveillance
Computer forensics tools	Recovering deleted data and piecing together a history of activity
Content filtering for e-mail	Detecting and filtering out unacceptable content
Content filtering for web browsing	Detecting and filtering out unacceptable content
Cryptographic hardware	Providing high-performance cryptographic processing, high security key storage, secure time source, random number generation for key management, tamper-resistant enclosures
Cryptographic software toolkits	Run-time libraries for data encryption, authentication, digital signatures and certificate processing
Data back-up management systems	Copying and storage management and restoration to a previous business position

8.4 Identities, Functions, Actions and ACLs

Please refer to Chapter 3, Figure 3.3, the SABSA Matrix, Component layer, People column, where you will see a cell entitled Identities, Functions, Actions and ACLs. This section discusses the main functional security protocol standards and their application. It is focused on the web services standards that are currently being used to build the infrastructure for digital business.

8.4.1 Web Services

The term web services refer to modular component functions that can be integrated to form the building blocks of web applications.

The consumer of a web service can be written in any language and supported on any platform. The only constraint is that it must comply with the XML/JSON message interface to the web service.

The listener is a service-agnostic component that simply receives and sends XML/JSON messages. It does not interpret the messages and therefore is generic to all web services. The business façade interprets the messages and is thus specific to the web service. It translates messages between the external XML/JSON message interface and the internal interface to the business logic. The internal interface is based on whatever existing middleware architecture you happen to have to integrate your existing business applications. To implement a web services architecture, you need the following components:

1. A standard way to represent data (XML representation of data and an XML schema
2. A common, extensible, message format
3. A common, extensible, service-description language
4. A way to discover services located on a particular web site
5. A way to discover service providers

8.4.2 XML Schema

You were introduced earlier to the idea of XML as a meta-language, used to create new customized languages for specific applications. In creating such a new application language there must be some agreement about shared vocabulary and processing rules since you can create any XML documents you want if the rules of XML are adhered to. This set of common definitions is created through XML schema. The schema defines the structure, content, syntax and semantics of the XML documents to be used. It is written using the XML Schema Definition Language, and it allows the rules that have been devised by humans to be interpreted by machines. There are many XML Schemas already defined, for instance for EDI purposes, so check if one already exists which can be used, before developing your own.

8.4.3 Simple Object Access Protocol (SOAP)

This is a peer-to-peer message exchange protocol for use in a decentralized, distributed environment, providing many-to-many connectivity. Fundamentally it is a stateless, one-way message paradigm, but it can be used as a building block to create application-specific, state-dependent message exchanges such as to request/response pairs of messages. It is formally specified as an XML information set.

The message structure is like that of most other conventional message protocols, with a message header and message body. The header contains routing and processing information and the body contains the end-to-end payload content sent by the originator and intended for the final recipient.

8.4.4 Web Services Security and Trust

As a part of the overall XML/JSON-based web services development, there are several specific standards addressing security and trust in this environment. At the time of writing this is a fast-moving area of development and so quite possibly when you read this thing will have changed considerably. In view of that, the detail is kept to a minimum here. You will need to go and research the most up-to-date position.

8.4.5 XML Encryption

XML Encryption17 is an open standard that defines techniques for using XML to represent encrypted XML and other data. It permits encryption of an entire XML document, of the specific elements or just of specific element content. Any encryption algorithm can be used, specified in an <EncryptionMethod> element.

8.4.6 XML Signature

XML Signature (XML-SIG) is an open standard for creating an XML-based digital signature for data of any type, including XML data. The signed data may reside within the same XML document as the signature or it may be located elsewhere. The signature itself is in the form of a short digest, which may optionally be encrypted as well. Any signature algorithm can be used, specified in an <SignatureMethod> element of the signature data.

8.4.7 SOAP Extensions: Digital Signature

This extension to the SOAP standard provides an open-standard method for digitally signing SOAP messages. It is based either on XML Signature or on other non-XML signature standards if required.

8.4.8 Security Assertion Markup Language (SAML)

SAML is specifically designed to create security assertions. It provides a framework for exchanging security information between business partners over the Internet. The key benefits of SAML are:

1. It is an open standard, designed to work with any industry-standard transport protocol
2. It provides inter-operability between end users, service providers and brokers
3. It provides single-sign-on across multiple web sites under multiple ownership
4. It is integrated into cloud environments

8.4.9 XML Benefits

The business benefits of using XML and its derivatives are that it:

1. Enables international media-independent electronic publishing
2. Saves businesses money by enabling the use of inexpensive off-the-shelf tools to process data
3. Saves training and development costs by having a single format with a wide range of uses
4. Increases reliability, because user agents can automate more processing of documents, they receive
5. Provides a foundation for the Semantic Web, enabling a whole new level of interoperability and information exchange
6. Encourages industries to define platform-independent protocols for the exchange of data, including electronic commerce
7. Allows people to display information the way they want it, under style sheet control
8. Enables long-term re-use of data, with no lock-in to proprietary tools or undocumented formats.

8.4.10 XML Security Architecture Issues

The main attacks against XML-based systems are by means of executable malicious code inserted into XML documents. To counter this type of attack several architectural approaches are needed:

1. All XML documents must be signed to detect unauthorized changes

- XML tunneling through conventional network-level firewalls needs to be controlled and managed by means of XML application-level firewalls

The XML family provides for a rich combination of security mechanisms embedded within the various protocols, but if application architects and programmers fail to design for an appropriate set of security services to be implemented through these mechanisms, then there will be no security advantage.

8.5 Processes, Nodes, Addresses and Protocols

Please refer to Chapter 3, Figure 3.3, the SABSA Matrix, Component layer, Location column, where you will see a cell entitled processes, nodes, addresses and protocols. This section describes some more security-related protocols and describes how these fit into the hierarchical protocol stack.

8.5.1 Protocol Stack

Those protocols that provide application security have been discussed in the previous section. This current section is looking at the component level of the network infrastructure, and so discusses the lower-level protocols. The processes in this infrastructure stack are always client and server processes, and these are all client-server protocols.

8.5.2 Hypertext Transfer Protocol (HTTP/S)

HTTP is the underlying protocol for communicating between a web client (browser) and a web server. It has been a huge success for two reasons:

1. Simplicity – the web architecture is simple for all participants
2. Ubiquity – by setting up a web server you automatically join a global community
3. Security – With encryption applied the communication is secure

All the specialist security services protocols that are discussed in the previous section (XML Encryption, XML Signature, SOAP Extensions, SAML) rely on a transport protocol to carry them between client and server and HTTP is the basic workhorse for this task. The later versions of HTTP, HTTP/2 and HTTP/3 come with additional security features built in, as well as increased speed of communication.

8.5.3 TLS

Secure Sockets Layer was developed by Netscape and as such is a proprietary protocol. However, it has been published and is widely implemented by many vendors. SSL version 3 was published in November 1996 as an Internet Draft.

Transport Layer Security (TLS) is the official non-vendor protocol from the IETF that replaces SSL and is specified in RFC 2246, 1999. There is an IETF working group focused on developing TLS. The latest version of TLS is 1.3, released at the end of 2018, but it has not been widely deployed at the time of this writing.

8.5.4 IPsec

IPsec is a large group of standards describing how cryptographic security is integrated into the IP packet layer. There are two basic approaches: Encapsulating Security Payload (ESP) and IP Authentication Header (AH). The key management is handled by the Internet Key Exchange protocol (IKEv1,

IKEv2). There are also two modes of applying the encryption, transport mode and tunnel mode.

Transport mode reveals the original IP source and destination addresses, whereas tunnel mode hides them and reveals only the addresses of the IPsec gateway. There is no mandatory encryption algorithm specified in the standards. Vendor support for certain algorithms is the only real constraint. The authentication is achieved by the use of keyed hashing (HMAC).SHA-2 and SHA-256-512 is the currently supported hashing algorithms. The IPsec standards are under continuous development by the IPsec working group of the IETF. IPsec is a technology for providing network security and is not a technology for providing application security.

8.5.5 DNSSEC

To complement the security mechanisms within IPsec it is also necessary to secure the DNS lookups, otherwise, the overall security of the network can be compromised beyond the control of IPsec. To achieve this there is an extension to DNS defined in RFC 4033 called DNSSEC.

DNSSEC applies digital signatures (using RSA or DSA) to authenticate DNS requests and responses. It also provides authenticated storage and distribution of public keys used for its own purposes and for other network-level secured protocols. DNSSEC does not provide confidentiality for DNS traffic, keep that in mind.

8.6 Security Step-Timing and Sequencing

Please refer to Chapter 3, Figure 3.3, the SABSA Matrix, Component layer, Time column, where you will see a cell entitled Security Step Timing and Sequencing. This section expands a little further on this subject.

The timing and sequencing of security steps are primarily driven by business requirements, like user expectations, business deadlines, volume throughput requirements and so on. However, in this cell of the SABSA Matrix, you must deal with the nuts and bolts of how that timing and sequencing is to be achieved in practice. Much of this will depend upon the performance and efficiency of the various components that you have assembled to build the architecture.

There is a detailed design task to be executed that focuses on how each component performs, fits into and how the timing of their individual operations can be interlocked to deliver the ultimate business requirements.

Look back to the case study in Chapter 5 under the section headed Time Performance Issues, which discusses the design of a process that involves decryption of database records. This will help you to understand the type of issue that can face you. In developing your detailed designs at the component layer, it is the interaction between components that you must consider. The actual sequencing of component operations may be different from the logical business flow, simply to gain better performance efficiency.

8.7 To Summarize: Component Security Architecture

One of the critical factors for a security architecture success at the component level is compatibility, consistency and interoperability between the various components. This is achieved through standards, international, national and industry sector standards.

Components exchange data and hence one of the most critical areas for standardization is the data structures and the protocols used to make the exchanges. Fundamental syntax standards such as ASN.1 and XML are essential to the construction of higher-level protocol standards.

New standards are being developed and published all the time, making this area somewhat of a moving target, but none the less important. To be aware of the most recent developments and of upcoming new developments you need to carry out up-to-date research on the various standards-making bodies and the individual vendors who are members of those bodies.

The security products and tools that are marketed by the vendor community fall into a few generic categories, each with a common set of features and functions. The chapter contains a high-level taxonomy of these products to help you to understand which components you might want to include or already y in place, at this level of your enterprise security architecture. However, the list cannot be exhaustive, and new products and tools are being launched all the time, so once again, up-to-date research is called for when you are selecting your components.

Many of the standards to which security products and tools comply are standard communications protocols at various levels. It is important to understand the functionality of each of these protocols – what they can do and what they cannot do and also to understand their relative positioning within the hierarchical protocol stack, which in turn governs the functionality and business benefits that they can bring you.

9

Security Policy Management

Security policy is the logical embodiment of the enterprise business requirements for security and control. It can therefore be something that, once determined, is a key driver of the operational security management program. This chapter looks at various aspects of security policy and how to manage it. In this chapter you will learn about:

1. Security policy as the logical model of your business requirements
2. How to use a security policy to develop a strong security culture
3. How to use risk assessment as the means to select the appropriate level of security policy
4. How to construct a hierarchical security policy architecture that is aligned with the layers of the SABSA Model
5. How to set up an organizational structure that supports the creation, implementation and management of security policy
6. How to manage security policy in an environment of outsourced technical services, like the cloud.

9.1 The Meaning of Security Policy

Please refer to Chapter 3, Figure 3.3, the SABSA Matrix, Logical layer, Motivation column, where you will see a cell entitled Security Policies. Security Policies and Security Policy Architecture. In Chapter 7 there is a brief discussion about these two deliverables since they fit into the SABSA Matrix row that is addressed by that chapter. However, as is explained in Chapter 7, the process of policymaking and policy management is more appropriately addressed as part of the operational security architecture and hence is the subject of this chapter.

A Theoretical View

There is a section in Chapter 6 with this same title. The theoretical basis for security policy comes from the concepts of security domain and security authority as described in Chapter 5 (Conceptual Security Architecture) and you may also find it useful to remind yourself of that discussion.

9.1.1 A Cultural View

On one level a security policy appears to just be words on a piece of paper, a purely physical document that lives in the filing cabinet somewhere in your office or on the Intranet. However, if your security policy is simply pieces of paper or screens with words on them, it has no value to the organization and you may as well discard it.

Security policy must be a living, breathing thing, regularly maintained as well. It represents a culture that exists in the organization. It describes the way in which people behave when doing their work. It is a mindset that has been accepted and bought into at all levels of the organization.

Security policy is not something that you can lift from a textbook, hence the absence of an example in this book. Security policy is a statement of business requirements for security, translated into a logical structure that can be consistently applied, monitored and measured. Your security policy should be unique to your business and fit the realities that you face in your particular business area. It communicates the intentions of your management team for managing risk and enforcing security in your organization.

9.1.2 Structuring the Content of a Security Policy

Through the long experience of working in a wide range of client organizations, the authors have seen many ways of structuring security policies. Some approaches work better than others. So, what is the right way to approach the structure and content of a security policy? The critical driver for making your decisions about the structure and content of a security policy should be:

1. What is the purpose of this security policy?

If your only purpose is to get a tick in the box when the auditor comes around, then it doesn't really matter what approach you take, you just want some shelf-ware. However, it is unlikely that collectors of shelf-ware will be readers of this book, so I doubt that is what you want.

What you probably want from a policy is a means to influence the mindset and hence the behavior of certain people in your organization. So that is the place to begin. Define the community of people to whom this security policy is to be addressed. The key question is:

2. Whose behavior are you trying to influence?

If, when you try to define this community, it turns out to be too difficult, then you have chosen the wrong scope for your policy. Go back to the beginning and ask again, what identifiable groups of people do you want to influence in mindset and behavior? Make a list of these groups and people. Then for each group consider, what sort of security policy do you need to influence this group?

Remember the key message of the Rules for Influencing Opinion and Behavior, that you should confine your content to only subjects that you think are relevant to influencing the target audience. Do not include irrelevant material 'just because it's there'. Keep your message concise and to the point. Focus on the objective, to make them think and behave differently.

9.1.3 Policy Hierarchy and Architecture

Given the principles set out previously, you will need policies aimed at different groups of people. One policy, the top-level corporate security policy, is aimed at everyone. This policy is short and high-level. It is issued with the authority of the chief executive officer and should carry his or her signature. This corporate policy is a message from the CEO to everyone.

Below this corporate-level security policy there will be other more-detailed policies that are aimed at specific sub-groups of people, maybe even subsets of systems, depending on compliance needs, although some of these more detailed policies may also be applicable to everyone, like an Acceptable Use Policy. These secondary policies add more flesh to the bare bones of the corporate security policy.

It is not possible to prescribe here the set of policies that you will need, for the simple reason that I do not know the target groups or your business and because your target groups will differ from those of another reader. Instead, here is some guidance on how to arrange those policies into policy architecture and how to structure supporting documentation to help the implementation of the policies.

The top-level policy is designed to bring together all the common themes of operational risk management across all the operational risk management disciplines. There are policy statements that are applicable to all these disciplines and it is better not to repeat these same statements under different policy headings but to assemble them together into an integrated, overarching top-level policy. Such a top-level policy is addressed to all employees throughout the enterprise.

At the next layer, there are policies to diversify operational risk into its constituent disciplines. For the purposes of this book, only the three domains of information security, business continuity and physical security are considered.

The third level for policies refers to the certification authorities and registration authorities (PKI) that one would expect in an enterprise security architecture built around the concept of public key infrastructure and digital certificates. In such an environment, potentially there will be a few domains in which entities are registered and issued with digital certificates. The authorities that control these domains, the certification authorities and registration authorities, have policies stating their requirements for security within the domains.

At the fourth level is the place for specific infrastructure policies, including examples such as Anti-Virus Policy, Remote Access Policy, Network Security Policy, Platform Security Policies and Acceptable Use Policy. Also at this level are the individual business applications policies. These might be divided by a line of business or by individual business application or both.

The decision on how to split up these policies will be driven by the granularity and variability of business risk across these domains. There is a further detailed discussion of this topic later in the chapter under the headings System Classification and Application System Security Policies.

Below all these various policies are several layers of supporting documents. These are various levels of detailed documents that help you to know how to implement policy at the nuts-and-bolts level, such as password syntax standards and how often a password should be changed.

At the fifth level of the policy hierarchy come the security rules, practices and procedures. All the policies must be aligned with the logical security architecture, as indicated in the introductory paragraphs of this chapter. However, the security rules, practices and procedures are aligned with the physical security architecture.

At the sixth level of the policy hierarchy, the security standards are aligned with the component security architecture. These standards include both external standards, international, national and industry sector standards, and internal standards. Finally, at the base of the hierarchy is a layer where you can place documents that provide implementation guidelines, where these are appropriate. This layer is likely to be sparsely populated because these guidelines are needed only in certain circumstances. Implementation guidelines are part of the operational security architecture at the Component layer (see Chapter 3, Table 3.2), because they give you advice about how to use or implement certain tools or products.

9.1.4 Corporate Security Policy

Now take another look now at the highest level of policy, the corporate-level policies. These remarks may apply to any of the policies referred to in the top two layers of the policy architecture. ISO/IEC 17799 sets out the objectives for a corporate information security policy as being:

1. 'To provide management direction and support for information security. Top management should set a clear direction and demonstrate their support for and commitment to information security through the issue of an information security policy across the organization'.

Your corporate level policies should address the following points:

1. A policy at this level should come from the most senior level of management.
2. To get such senior sponsorship, you will need to invest many months in developing the policy, mostly in a process of wide consultation across the organization to ensure that it represents a consensus view and is not. just the personal opinions of an individual.
3. The policy should refer to business risk management as the primary driver.
4. The need for security and risk management should be related to the overall business goals of the organization.
5. The policy should provide a strong mandate, instructing the management at all levels and the employees at large to behave in certain ways and holding them accountable for doing so.
6. The policy should formally delegate responsibility with phrases like 'divisional directors are responsible for...' 'line managers are responsible for...' and 'all staff are responsible for. . .'
7. It should specifically mention compliance with laws, regulations and contracts.
8. It should refer to a reporting process for suspected security incidents.
9. It should mention the need for education and training and the provision of a center of expertise to provide internal advice and support on security matters.
10. It should refer to other documents where more detail can be found.

9.1.5 Policy Principles

There are several principles that may help you in formulating security policies. These are not presented as something to which you must slavishly adhere. Indeed, some of these principles may be counter to your intentions in some circumstances. Look on them merely as a resource from which you can draw inspiration.

1. **Least privilege principle** – Users and system processes should be given the least authority and minimum access to resources required to accomplish a given task
2. **Accountability principle** – All significant system and process events should be traceable to the initiator.
3. **Minimum dependence on secrecy principle** – Controls should still be effective even if an opponent knows of their existence and knows their mode of operation.
4. **Control automation principle** – Wherever possible, automatic controls should be used rather than controls that depend on human vigilance and human behavior.
5. **Resiliency principle** – Systems should be designed and managed so that in the event of breakdown or compromise the least possible damage and inconvenience are caused.
6. **Defense in depth principle** – Controls should be layered such that if one layer of control should fail, there is another different type of control at the next layer that will prevent a security breach.
7. **Approved exception principle** – Policy exceptions should always have management approval.
8. **Secure emergency override principle** – Controls must only be bypassed in predetermined and secure ways. Systems are at their most vulnerable when normal controls are removed for emergency maintenance or other similar reasons. There should always be procedures and controls to minimize the level of risk in these circumstances.
9. **Auditability principle** – It must be possible for an independent expert to verify that thesystem conforms to the security policy. A necessary, but not totally sufficient condition for this is that the system must be able to record security-related events in a tamper-resistant audit log.

9.1.6 More About the Least-Privilege Principle

The least-privilege principle is a longstanding security policy principle that you will find stated in any serious text on information security and without a doubt, it is an important principle in all circumstances. For example, at the level of platform security, it works very well.

The principles here require that an organization can use existing business information to create new business knowledge (refer to Table 9.1 in Chapter 6 for the clear distinction between these two terms). The whole basis of data mining is developed from this idea. Thus, for a user to be able to create new business knowledge, access to all business information may be required.

One way to view this issue, which does not then violate this least-privilege policy principle, is that those whose job function is to be a knowledge worker require access to all business information all the time and this is just a special case of the least-privilege principle. Whichever way you view this, you should be aware of this issue and ensure that your approach to information security policy does not inhibit genuine business activity related to knowledge.

9.1.7 Information Classification

One way to approach the security policy and its implementation is to classify information into one of several classifications, each of which has an associated security policy. Thus, once classified, that information must be handled under the terms of the associated security policy. This can be a daunting task, but the overall security benefits, with regards to DLP and rights management are enormous!

The concept of information classification is one that has been developed over many centuries in military and government organizations. The focus has traditionally been on differing requirements for secrecy, depending upon the information to be protected. Thus, it is usual to define several levels of classification, such as top secret, secret, confidential and restricted, leaving everything else unclassified by default. Each document is classified at one of these levels. Each object is also protectively marked to ensure that its classification is obvious and can be used to make decisions as to how it should be handled, according to the relevant security policy. A lot of tools out there promise to help with the classification process, but none of the ones I am familiar with can automate this to the degree that is often promised. So, even with a tool, expect to use a significant amount of time/resources in getting this into production.

The people who might gain access to documents are given security clearance at one of these levels. Thus, someone who is cleared to secret level can read any document classified up to and including secret but is not allowed to read documents classified as top secret. This matching of subject clearance levels to object classifications is at the heart of multi-level secure systems (MLS).

MLS are computer systems, usually in military or government establishments, which simultaneously handle objects at multiple levels of classification and subjects at multiple clearance levels. The Bell-LaPadula model is a formal description of access control in MLS.

The idea has now been applied to data integrity. Differing levels of integrity protection can be defined, such as 'sensitive for personal health and safety', 'sensitive for business mission'; 'sensitive for business function' and by default, everything else is 'non-sensitive'.

The Biba model formally describes a MLS managing multiple levels of integrity. The basic rules are summarized as 'no write to a higher level' and 'no read from a lower level', the opposite of the Bell-LaPadula model. Whilst these models are all well and good in theory, in practice they prove to be difficult to make operational and useful, especially in a civilian environment. Even in the military environment for which these models were developed, they pose many practical difficulties and they need to be applied with great care.

The situation can get worse. Once you move beyond the military and government arena into a commercial business, the emphasis is rarely on secrecy. Continuous availability and integrity are usually much more important business drivers. Just supposing you can justify a classification scheme on business grounds, you still have many operational problems. Each classification category needs rules that determine how it gets classified at that level, in accordance with the security policy. Different departments have different names and classifications of documents. Everyone who creates documents or objects that might be at that level needs to know those rules and needs to apply them rigorously. All documents need to be given a protective marking. Then the people who may receive the documents need to know how to handle them. Someone cleared to a high level may have to manage several levels of classification, must know the different handling procedures for each and must apply them rigorously.

This is all a lot of hard work. It can have an adverse effect on efficiency and upon the flow of information in the business and it has a high cost. There is another problem. Because it is unusual for items to be downgraded in their classification, information tends to drift upwards in the hierarchy, usually attaining a higher classification than it really needs. In conclusion, if you already have or are considering implementing an information classification scheme, think carefully about the issues:

1. Do you really need this for business purposes?
2. What added value will it bring, and at what cost?
3. How will it work in practice?
4. How much user training and support will it require?
5. Can it be made simpler and easier to operate?
6. Will it really be used or will it soon fall into disrepute and disuse?
7. Does it really deliver any benefit?
8. Should you really reject this approach altogether?

An interesting aside to note is that ISO/IEC 17799 mentions information classification as an approach but does not detail any of the discussions you have seen above about the value or difficulty of classifying objects and handling classifications.

9.1.8 System Classification

A much more useful way to approach classification to manage security policy might be to classify systems or applications according to the level of business risk revealed through a risk assessment. The reason to do this is to set a security policy for each level of risk and to associate each system with a specific control regime, a standard set of controls that must be applied to protect the application system according to the level of risk as specified in the associated security policy.

You can diversify this to another level by having the risk level stated for each of a small number of Business Attributes (such as confidential, integrity protected and available) because the standard control regime for each will be slightly different. Thus, a system may be classified as high-risk for available, medium-risk for integrity-protected and low-risk for confidential. By reference to the security policies and control standards for each of these risk categories you can see what the complete set of controls should be to protect the system to an appropriate level of security as specified in the security policies. The benefits of this approach are:

1. It provides a method of ensuring a consistent, standardized approach to enforcing security policy across multiple systems and applications.
2. Although there are many, many application systems, only a limited number of security policies are maintained to apply to those systems.
3. It ensures that the level of security and control on each application system is matched to the business risk for that system.

4. It is completely transparent to the user community, which does not have to get involved with any decisions about how to handle a specific document because the system managers apply the security policy control regime at the system level, not the user level.
5. It makes the job of security auditing of systems much simpler because it removes subjective judgment and introduces an objective set of criteria for the audit.

9.1.9 CA and RA Security Policies

The security policies for certification authorities and registration authorities are specifically focused upon managing the authorization of registrants and embodying that authorization in a logical format in a digital certificate in PKI-based enterprise security architecture. CA and RA security policies must cover the following issues:

1. Processes used to register entities and issue them with certificates
2. Security management for the CA and RA ICT systems
3. Validity periods for an entity encryption public key certificate
4. Validity periods for an entity-signing key
5. Validity periods for an entity verification public key certificate (which must be greater than or equal to lifetime for the entity-signing key
6. Validity periods for lower-level CA signing keys
7. Lifetimes for certificate revocation list

9.2 Application System Security Policies

If you adopt the approach described earlier in this chapter whereby each application system is risk assessed against a few Business Attributes then you will need an application system security policy for each risk and attribute category, nine in this case, as shown in Figure 9.1.

This small number of application system security policies can be applied to many application systems, making the process of policymaking and policy management efficient and ensuring consistent risk management practices across all applications. If this approach seems to not provide the granularity

	Low Risk	Medium Risk	High Risk
Confidentiality	Policy 1	Policy 2	Policy 3
Integrity	Policy 4	Policy 5	Policy 6
Availability	Policy 7	Policy 8	Policy 9

Figure 9.1   Risk Attribute table

that you require for business purposes, then you can of course have an individual risk assessment for each application, based upon the entire Business Attribute Profile for the line of business supported by that application. This will be less efficient in terms of effort but will provide a much more finely tuned security policy.

Another possibility is to have a set of security policies for each risk level paired with each line of business. This provides another efficiency gain by limiting the total number of policies in use.

The issues to be addressed in an application system security policy are concerned with the Six As of application security, as described in Chapter 5 in the section entitled Security Services in the Application Layer:

1. Authorization
2. Authentication
3. Access control
4. Audit
5. Administration
6. Application-to-application communications security

9.3 Platform Security Policies

A platform is a combined hardware box and its operating system, possibly virtualized. It may be used to host a single application or more often, multiple applications.

One important distinguishing factor regarding security policy is the operating system, but there may also be key aspects of the hardware that will also drive the security policy. There is an almost infinite number of combinations of hardware and operating system that you might need to support, but hopefully, you have an ICT strategy that limits the number of platform types to improve the efficiency of administration and support.

You will need to categorize your platforms into several types that are reasonably homogenous across each type. Then you will need to develop a security policy for each of these platform types. However, there are more issues to consider:

1. Should the security policy for the platform be independent of the risk level and security policy for the application being hosted? (OS only?)
2. If the application risk level is to be considered, in the case of multiple applications being hosted on the same platform, will the platform security policy be the highest applicable?

The more granularities you introduce, the more complex and the less efficient becomes the process of managing the security policies. There are real cost/benefit trade-offs to be made in deciding the level of granularity at which security policy will be formulated.

9.4 Network Security Policies

You will almost certainly need to develop a security policy that governs the entire enterprise network domain and applies to all parts of it. You may also need to develop a few network sub-domain security policies, depending upon the nature of your business and the degree of separation that you require between these sub-domains, PCI DSS for payment card data for instance. Once again, efficiency can be traded against complexity and granularity, but in the networking domain, this is unlikely to be as great an issue as it is with applications or platforms.

Where you have implemented VPNs, these will need to be governed by a security policy, perhaps one for all VPNs, but more likely based on where the VPNs connect, or the application is accessed via VPN. Firewalls also need a security policy that gets converted into rules at the physical layer for configuring the firewall. The issues that your network security policies will need to address and the strategic principles of network security are described in Chapter 10 in the section entitled Security Services in the Network Layer.

9.5 Other Infrastructure Security Policies

There are other aspects of ICT infrastructure that will need specific security policies. These may include:

1. Anti-virus and other malware policy
2. Remote access policy
3. Acceptable use policy
4. User authentication policy
5. Database security policy
6. Data management security policy
7. Security management services security policy
8. Directory service security policy.

9.6 Security Organization and Responsibilities

Who within the enterprise is responsible for information security? The correct simple answer is 'everyone', but the full answer is more complicated than that. Clearly, there is a hierarchy of responsibilities in play here, which implies that there is an organizational structure dedicated to information security management, including security policymaking. What should that structure look like?

It absolutely must start at the very top, with the company board and the chairman. At this level, there is a responsibility to the shareholders to ensure that their investments are being properly managed. This falls under the general umbrella of corporate governance and includes the setting of high-level risk management policies.

At the next level are the CEO and the other members of the C-Level management team who report to the CEO. They must execute the directions from the board and take executive responsibility for all aspects of corporate risk management. This includes the management of information security. To make this possible this executive team must set up a specific organizational framework to manage all aspects of corporate security in general and information security.

The director in charge7 of information security could be any member of the senior management team and its role must be part of senior management and not attached to the CIO role at a lower level.

So which direction should it be? If this is really going to work, then the director should be genuinely the flag-bearer for the information security cause. However, there is one director who should be avoided in making the choice of the director in charge of information security, the IT director. If this responsibility is placed with IT, then it sends the wrong message. It is essential that information security be understood by everyone to be a business issue and not an ICT issue. The more involvement there is from the IT department, the fewer people will understand the true business importance surrounding information security.

Another clear recommendation from ISO/IEC 17799 is the formation of a management steering group at which major security policy and strategy issues get discussed. This will provide clear direction and visible management support for security initiatives.

The terms of reference and power of this steering committee are critical to its success. If it is not vested with any real power, then the senior people who are supposed to attend will always be busy doing something else and will always delegate someone else to go to the meeting instead. You will then finish up with a junior committee that is a talking shop and adds little value to the process of managing enterprise information security. You must create an environment in which the senior members of the steering group regard it as important to attend because important decisions will be taken that they want to influence.

There are other positions of authority and leadership that need to be defined. In an organization of any real size, like a CISO but the CISO can no longer be expected to handle all areas since some of it will be dependent on external regulation. A Data Protection Officer (DPO) was defined as part of the EU GDPR regulation.

There are a few other roles that need to be fulfilled. How these roles are integrated into the organizational framework will depend on the enterprise and how it is organized, as well as national and regional regulations. Here are some of the roles that can be identified:

1. Security administration
2. User support
3. User awareness and management awareness development
4. Specific education, training, and skills development
5. Security audit – Internal/External
6. System security administration: configuring IT systems.

There needs to be a co-ordination framework for these various people who may be widely distributed across the organization, perhaps even globally distributed. You may need another level of the committee below the steering group, called the Information Security Liaison Group. This is an informal group, probably led and chaired by the CISO, the purpose of which is to communicate with this distributed community of people who have an interest in information security. The company board is responsible for:

1. Corporate governance
2. Setting goals and expectations for risk management

The Executive Management Team is responsible for:

1. Ratifying policies approved by the Information Security Management Steering Group
2. Approving budgets for information security initiatives and programs

The Information Security Management Steering Group is responsible for:

1. Reviewing and approving corporate security policy and subordinate policies

2. Approving and supporting major initiatives to develop the enterprise information security program
3. Developing and submitting to the Executive Team for approval all major budgets for information security-related activities
4. Monitoring major information security-related threats to the enterprise business
5. Monitoring and reviewing significant security incidents.

The Information Security Liaison Group is responsible for:

1. Providing a communications channel to listen to views, ideas and inputs from all those with operational responsibilities
2. Providing a dissemination channel to communicate new policies, processes, and methodologies
3. To find a resolution for operational problems or to escalate these if an immediate resolution is not possible.

Line management is responsible for:

1. Owning the corporate information assets on behalf of the enterprise
2. Ensuring compliance with corporate information security policies, practices, procedures and standards throughout the span of control
3. Developing awareness of information security issues and a strong culture of information security management throughout the span of control

9.7 Security Culture Development

Information security is everyone's responsibility. However, this can only work if everyone understands that and knows how to discharge that responsibility. Thus, it is important to develop a strong information security culture right across the organization.

The first key component of this culture is a corporate information security policy statement signed off at the most senior level of executive management, the CEO and possibly the board as well. Everyone must see this policy as the lynchpin of the information security stance of the organization.

Having the policy written down is not enough. A real policy is one that lives and breathes in every action that anyone in the organization takes. That takes strong awareness, and strong awareness can be developed only by constantly bringing the issue to the attention of people. Even if you raise awareness to a high level at a given point in time, if you neglect it after that, the awareness decays and disappears. People forget. People leave and new people arrive. So, you need a campaign that will constantly renew the awareness level by a continual series of reminders that never stops.

One key driver of awareness and commitment throughout the organization is the example set by management. Managers at all levels must set a good example by always abiding by the policy and by pointing out compliant behavior as well as non-compliant behavior. Managers must be always on the case, never letting the issue of security slip, especially when things have gone wrong.

On the other hand, a culture of blame will only encourage people to cover up mistakes. Much better to have a 'no-blame' culture in which everyone is encouraged to admit their mistakes, to own them on a team basis even though an individual may have made a mistake, as we all do on a regular basis!

However, there are circumstances where blame and retribution are appropriate. These are cases where an individual, either with deliberate malice or with reckless negligence, has caused a major loss event which should have been prevented had that individual behaved properly. It is a matter of scale. Sometimes people are so reckless, malicious or incompetent that they need to be punished and possibly removed. It is therefore essential to have in place an appropriate disciplinary process as part of the overall human resources management process.

Regular consistent reporting of security incidents is an important part of cultural development. It works best in the open, no-blame culture described above, because there are no inhibitions to reporting an incident in those circumstances. In these circumstances the number and type of incidents can sometimes be used as a metric of success.

Another aspect of security culture development is the education and training program. There are several different types of education and training that should be undertaken, according to the specific business needs of the enterprise. These include:

1. Brief induction training for all new employees to ensure that they are fully aware of the security policy
2. Specific technical training for anyone whose job includes a technical activity relevant to maintaining the enterprise security stance.
3. Professional career development for those with specific information security responsibilities

4. Short courses on various aspects of information security, including new issues as they arise.

I know that there are companies and organizations out there, with the attitude that if they give their employees education and certifications, then they risk them leaving for greener pastures elsewhere. Yes, that is a risk, but what do you have running your critical business systems if you do not?

9.8 Outsourcing Strategy and Policy Management

There are three main issues to be addressed regarding outsourcing strategy:

1. The security of outsourced IT services
2. The security of business information handled by service provider
3. The outsourcing of operational security management services.

Many organizations outsource activities that are regarded as non-core business or move them to a cloud environment. For an enterprise that is not an IT services company, the operational management of ICT is often a prime candidate for such outsourcing. This includes both standard data center operations and call center operations based on computer-telephony integration. The management of the security of the ICT systems and the business information they process must be addressed as a part of the outsourcing contract.

Non-IT business activities are also frequently outsourced and in this area, there is currently a huge growth, motivated mainly by a desire to reduce headcount and to take advantage of low-cost labor resources in offshore locations. Those activities that are components of business processes, but which do not add real value to the process are the main candidates for outsourcing. For example, an insurance company might outsource a purely administrative function such as claims handling9, but in doing so it hands over to a third-party huge amount of confidential, business-critical information. Have you thought about information security in the contract with the vendor? Are you allowed to audit against it? There are obvious information security management issues to be addressed here.

9.9 To Summarize

The theoretical basis for applying security policy is embedded within the definition of a security domain, which is a set of security elements subject to the same security policy, defined and enforced by a single security policy authority.

Security policy is the logical representation of the business requirements of the organization for mitigating risk and enforcing security and control and is defined according to the needs of the various business domains. These domains can be arranged into a hierarchical layered policy architecture model and below this, other layers of supporting documentation are defined, procedures and standards, aligned with the lower layers of the SABSA Model.

Security policy is much more than a mere set of statements, it represents the culture of the enterprise and describes how people approach their work and how they behave about security matters. For this reason, you should begin writing a security policy from the point of view that you want to influence the attitudes and behavior of a specific community of people. The policy should speak clearly to that group of people, inappropriate language and stick strictly to the messages that the group needs to hear. Finally, please remember that a policy document only has value, if it is being maintained by the organization, as the risk and threats are changing.

10

Security Architecture – Cisco & Microsoft

At no point in the previous chapters, did I mention any specific technology, except when I told you about this chapter? When you develop a security architecture, you should wait, if possible, before putting concrete technologies or vendors, into the plan. For the practical chapter in this book, I have chosen Cisco and Microsoft, simply because those are the technology vendors, I am most familiar with in my consulting work.

Before we get started on these practical implementation steps, I should give you a description of how I will be developing the security architecture. Fully describing a security architecture, even with 'only' two technology areas, would take several hundred pages, it could probably fill an entire book on its own. So, the individual layers of the SABSA model will be developed, more or less, in point form, albeit with more description of the individual concerns in each of the SABSA layers. This will give you insights into the kinds of questions that should be asked of the companies designing a security architecture based on the SABSA model.

10.1 Use Case

For us to ground the development of a security architecture with equipment and software from Cisco and Microsoft, in a concrete scenario I have created a fictitious company. Although fictitious it is a company that could exist in any country around the world and probably does.

Our company is named N7 and it has developed a nice sideline in providing IT business consulting to their clients. The N7 specialty is aligning an organizations IT environment, including Cloud, with the business needs and regulatory reality. See Table 10.1 for some more details on N7.

N7 has not developed a formal security architecture before but has seen the light after being made aware of SABSA. The case for the Cisco networking infrastructure is an advancing age of the hardware units used to run

Table 10.1   N7 Requirements

Item	Description
Regulations	N7 itself is not under any sort of regulatory pressure, however many of their clients must deal with various regulatory and compliance requirements, like HIPAA, PCI DSS, GDPR, ...
Compliance	No external pressure on compliance exists for N7, but due to the regulatory compliance pressure that the N7 clients are experiencing, N7 has decided to implement and certify against ISO27001 for security management and ISO27701 for privacy management to protect the data that they are collecting from and developing for their clients
Business Strategy	To remain relevant for their customers, N7 is continually keeping up to date with new and upcoming regulations and industry compliance issues. The strategy is for N7 to be able to implement and advice customers on the new and upcoming regulations and compliance in the different business areas of their customers. Most of the N7 customers comes from: 1. Healthcare 2. Gambling industries 3. Financial institutions 4. Investment banking All of which are heavily regulated around the world.
Infrastructure	The N7 infrastructure is based on technology from Cisco, Microsoft, Microsoft Cloud (M365, Azure, Dynamics 365,) VMW are and a few custom developed applications, built on .NET
IT Strategy	The strategy for N7, on the IT side, is to move as much as possible into the Azure cloud and to use the Azure cloud as the application platform for new development efforts. N7 is already using M365 and Dynamics 365, so the cloud is not a new experience for N7. Some applications will be kept on-premises, but the majority of the existing server and application infrastructure will be moved to Azure.

the networking infrastructure. These units have now reached an age, where a hardware refresh is needed, to keep running hardware that is under support from Cisco. Something that every company will experience on a regular basis.

Since a lot of the essential infrastructure will be in flux for a while, the N7 leadership has decided that part of the server infrastructure should be moved to Azure as part of the exercise. The decision is made because the Azure cloud has shown significant promise in the various POC's that N7 has conducted on Azure, making Azure a core part of the IT strategy for N7.

As part of the hardware refresh and Azure move an N7 has decided to utilize the SABSA methodology for developing the parts of the security architecture where the networking layer and the Azure Infrastructure and Application layer are the primary factors in the SABSA model.

Any modern network equipment vendor will have devices that are no longer limited strictly to the networking layer. This goes especially for a vendor like Cisco, which has devices like the Web Security Appliance (WSA), that are working at the application layer. This brings me to an important piece of advice! SABSA is a set of guidance, not hard rules. In a case like Cisco and Azure, where the equipment/services can fit into multiple different layers, we will have to decide how to fit the equipment and service design into our SABSA model.

If at any point you are feeling constricted by SABSA and then feel free to adapt SABSA to the situation you are in! SABSA cannot be expected to cater to every situation and business environment, so use SABSA as a guidance, not a set of rules.

That being said, the examples in this chapter are examples of how I would do it, which should in no way be interpreted as the 'one and only way'! I make no claim to the philosopher's stone, with regards to security architecture.

10.2 SABSA

Before we begin the process of architecting our hardware refresh for the networking equipment, let's refresh the SABSA layers in Table 10.2

Table 10.2 SABSA Layers refresh

SABSA Layer	Description
Contextual	Business driver development, business risk assessment, service management, relationship management, point-of-supply management and performance management
Conceptual	Developing the Business Attributes Profile, developing operational risk management objectives through risk assessment, service delivery planning, defining service management roles, responsibilities, liabilities and cultural values, service portfolio management, planning and maintaining the service catalogue and managing service performance criteria and targets.
Logical	Asset management, policy management, service delivery management, service customer support, service catalogue management and service evaluation management.
Physical	Asset security and protection, operational risk data collection, operations management, user support, service resources protection and service performance data collection.
Component	Tool protection, operational risk management tools, tool deployment, personnel deployment, security management tools and service monitoring tools.

These next sections of this chapter will be divided into the layers mentioned in Table 10.2, where I will make some of the points that I feel will create a successful security architecture for N7. There will be one glaring omission in the chapter on practical security architecture development. How to implement the security architecture in the organization! This is by choice since every business or organization will have different challenges with regard to the implementation of changes.

Depending on the culture of the business/organization, changes can be challenging to implement into existing business processes. As a consultant or employee, of the business, you should spend some time developing insight, into the culture of the organization and developing the change process to cater to the cultural intricacies present in the company/organization. A difficult proposition to be sure, but a very important part of any change implementation. Let's get started, first up, the contextual layer of SABSA.

10.3 Contextual

First up, let's look at the drivers for N7. Based on the information in Table 10.1, we can expect the following to be true:

1. The Clients of N7 are highly regulated and in my experience, it takes some time to become a trusted partner in these industries. So, trust is a good word to apply as a driver for N7.
2. For any long-term relationship between N7 and their clients to develop, N7 must be seen as both a respected entity and a reputable entity, something that is especially important to the banking industry these years. So, the next two words to apply to N7 are Respected and Reputable.
3. Every industry around the world must deal with the external pressure for greener and more sustainable operations; N7 is no exception to this pressure, making sustainable another word to apply here. Sustainable is not only meant as a green parameter but it should also be seen as a parameter for the overall operations of N7 as well.

Based on what we have hare, the security architecture should be developed with an eye to:

1. Trust
2. Respect
3. Reputation
4. Sustainability.

At the contextual layer for N7, we do not necessarily need any more information before moving onto the conceptual layer. If this seems light to you, which is fine, the detail needed at the individual layers depends on the industry and the regulatory pressure they are under. For N7 there is no regulatory pressure for N7 themselves, all the pressure is aimed at their clients. Since N7 has a stated goal of following ISO 27001 and ISO 27701 we could add compliance as an additional word to describe the contextual layer for N7. But any good security architecture will be able to cater to the various standards out there, including the ISO ones.

Political considerations might dictate that compliance is part of the descriptive set of words during the design and development of a security architecture. So please feel free to add and remove words in the list when developing a security architecture for your own organization/company.

10.4 Conceptual

The contextual step was more of a broad list of objectives for the security architecture. With the conceptual layer, we are linking these broad objectives to the technological attributes needed on the plan for the security architecture.

Translating these high-level business requirements into IT objectives can be a daunting task for anybody! Involving the IT department in this step will, hopefully, bring them on-board with the development of overall security architecture. Also, they are the ones with the technological insight into the capabilities of the individual technologies in place. In this case, the IT department has already created a list that we can use to develop the attributes:

1. Internal and External information sharing
2. Enable Assurance Services
3. Strengthen Operational Service Management
4. Develop an enterprise-wide security architecture.

Depending on the situation and the complexity of the organization, this list can become quite long, but this step is the foundation for us to begin the development of the security attributes we need for the conceptual layer of the SABSA model. Some of these attributes could be from the list below:

1. Availability
2. Confidentiality
3. Access Control
4. Identity Management.

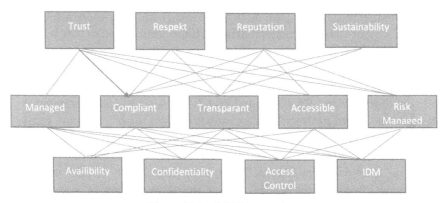

Figure 10.1   SABSA mappings

The next step here should be to map the relationships between the security attributes we have identified between the levels:

The three layers in Figure 10.1 are:

1. Enterprise Goals
2. Technology Objectives
3. Security attributes.

I am sure you can see from Figure 10.1, that these kinds of mappings can become quite complex! Non the less, this step in the SABSA process must not be underestimated or taken lightly.

10.5 Logical

Now we will put in a few more details into our N7 SABSA security architecture. At the logical layer, we are adding in the policy management for the hardware refresh of the Cisco networking infrastructure and move to Azure for N7. Let's first look at some of the IT objectives we have developed so far, to get a handle on the kinds of policies we will need to accommodate those. A list of possible policies for each of the technology areas can be found in Table 10.3.

You might be surprised by the presence of a networking automation policy in Table 10.3, but network automation is becoming more and more important in many network environments and Cisco is making a huge push for automation with their DevNet certifications. Automation in the network is also automation of the security running at the networking layer!

Table 10.3 Policy Options

Technology Area	Policies
Cisco	1. Network Access Policy
	2. Acceptable Use Policy
	3. Network Security Policy
	4. QoS Policy
	5. Network Automation Policy
	6. Network Governance Policy...
Azure	1. Application Security Policy
	2. Application Development Policy
	3. Access Policy
	4. Azure Cloud Security Policy
	5. Azure Governance Policy...

From Table 10.3, you can see that there are quite a few policies that we will need to create and implement as part of a SABSA security architecture, and the list is not even exhaustive. Add to that, that most of these policies will have to be developed in collaboration with different units within the organization and you can see the amount of work in the logical layer of the SABSA model is not to be underestimated.

You might be surprised by the lack of a compliance policy in table 10.3? Didn't I say that compliance was a parameter in the strategy for N7? I did. But trying to apply ISO standards at the same time you are developing a security architecture, will be too big of a task for anybody. If you go about creating and implementing a security architecture with an eye towards an ISO certification, that will be a much easier prospect and well-developed security architecture will align with the ISO standards very nicely.

10.6 Physical

The component layer is where we will begin to put the concrete technologies into our security architecture. The physical layer is where we add in the various security services that we need in our security architecture. By services, I am referring to the QoS, mentioned in table 10.3, but there are a plethora of services needed in any modern infrastructure to make it work and easy to manage.

Below is a list of some of the security services N7 will need in their SABSA security architecture:

1. Log Management

2. IDS/IPS
3. Access Management
4. User Support
5. Asset Management
6. Back-up/Restore...

Any of the services in the above list might be a separate unit within N7, one that is managing the back-up services for instance, as well as any incident response processes related to the back-up service. But, security services, are they not needed across the organization as a whole? Don't we need some insights into where these services are needed? Yes, we do!

The risk assessment that we did back in the contextual layer will provide us with the information we need to identify the areas, where security processes are necessary for the success of our security architecture. In my experience, designing these security services is not a challenge, although it can be laborious, but the operationalization of the services typically takes a lot of time to implement. This is another area, where the ability to communicate and collaborate in an organization will benefit the security architect.

10.7 Component

Now we get to put concrete technologies in our security architecture. Let's begin with the hardware refresh of the networking infrastructure, but first I am going to go of on a tangent here but stay with me.

With the increase and importance of cloud environments, the infrastructure we are using to access the cloud environments is becoming more and more critical, something that is often forgotten, when deciding on networking equipment. If we, like N7, have most of our infrastructure running in the cloud, then it does not matter if these clouds are up, if we cannot access the infrastructure running on them. If our local network is down, then we cannot use the cloud resources, this makes the local area networks a critical business resource, something that is often forgotten in the procurement and management of the networks.

10.7.1 Cisco

Back to the security architecture. So, what kind of hardware do we need from Cisco, considering my comment on the importance of the network above? Let's start with the basic infrastructure equipment:

1. Switches

2. Routers
3. Access Points
4. Software for the management and access to the network
5. Firewall(s)
6. IDS/IPS.

I will refrain from putting in concrete model numbers or software versions since they are less important than the overall idea behind the security architecture of the networking layer. Since we have already identified the network as a critical business resource, we must apply redundancy to each of the components in the hardware units we decide to put in.

1. Dual power supplies
2. Dual Supervisors (A Cisco network management component).

On top of this hardware and software, we will apply some of our security services, like patch management, configuration management and back-up. Do we need to back-up our networking equipment? Yes, we do. The configuration files running on the individual units need to be backed up in case of a unit failure. If we do not have the configuration files elsewhere, then we will have to reconfigure the unit manually, adding to the time it takes before that unit can be returned to production. Also, manually configuring something, adds to the risk of human mistakes!

This takes care of the local part of our networking infrastructure, which brings us to the Internet and WAN part of the equation. We need Internet if we are to have access to the cloud infrastructure and WAN services for integration with partners and vendors. Internet could be used for both of those needs, but in most cases these connections are separate. New technology has arrived making the management and administration of external network connections much easier. SDWAN.

With SDWAN, an abstraction layer has been pulled over the complexities of external network connections, like MPLS, VPN, Direct Access, ... SDWAN can even be used to optimize the external connections that are used for cloud access, making SDWAN an obvious choice for external network connections. Cisco is not the only vendor with SDWAN products in all fairness.

Here I am going to mention a specific piece of software from Cisco, namely DNA Center. Cisco DNA Center can be used to manage SDWAN and move the access control on the local network onto SDA or software-defined access. With DNA Center we can even manage and control the networking units on our local area network as well as the patching of these units. Central control, management and administration are the holy grail of networking and DNA center brings us very close to that reality, although DNA Center can be quite expensive, depending on the license type needed for the network.

Cisco comes with a LOT of equipment on the network and network security side of security architecture and depending on the needs of an organization and the architecture under development, some of the below technologies can make sense in each context:

1. Identity Services Engine (ISE) – used for granting/denying access to the network based on things like posture or location within the network
2. Cisco Umbrella – A DNS Security solution
3. VPN Technology in multiple different forms
4. Web Security
5. E-mail security…

10.7.2 Azure

Azure is a massive platform, with an almost infinite number of options and that is without even looking at the 3rd part options that can be used within Azure. The N7 strategy for Azure is to move part of the server infrastructure to Azure, making the IaaS offering one of the options that N7 needs. Moving to the IaaS offering in Azure does not release us from the responsibility for maintaining the patch level of the operating system and the applications running on top of the OS. That is the one thing we should keep in mind when migrating to an IaaS offering.

There are many reasons for moving to IaaS, one of them being easy of migration. We just need to move the virtual machines from our own datacenter to Azure and voila, we are running in the cloud. To really benefit from cloud migration, we should move to the cloud-native software offerings, like Azure SQL, Cosmos DB, Storage and so on. This will require more work, than just doing a lift and shift, but it will provide us with the benefit of having Microsoft do the maintenance and patching of the underlying infrastructure. We will just be responsible for the maintenance of our own applications and their code, releasing us from the security headache that is maintaining the Virtualization software, OS and applications running on those.

N7 is moving part of the server infrastructure to Azure, which will make the below areas the responsibility of N7:

1. OS Patching
2. Application Patching

3. Back-up
4. Antivirus...

The only thing we are really freeing ourselves from, is the hardware, virtualization, cooling and space needed for the servers, with IaaS in Azure. This should not be interpreted as IaaS being a bad choice, there are many circumstances where utilizing IaaS makes sense, like running SAP in Azure or moving a self-developed application to Azure, which would require significant recoding if it was to become a native cloud application.

Focusing on the IaaS for now, the security services needed for the IaaS deployment, are the exact same as was needed while these servers were on-premises:

1. Access Management
2. Patch Management
3. Back-up
4. Antivirus
5. Monitoring
6. Administration Management
7. Segmentation of the virtual networks.

So, there are a few different policies and procedures that need development and operationalization with moving to IaaS in Azure. Some of the ones we have been using on-premises can be adapted to Azure, assuming we had policies and procedures in our on-prem environment...?

Azure is a massive platform to gain access to as a customer and the great benefit for the customers is the ease with which you can get started with Azure. The great disadvantage is how easy it is to get started with Azure. By that, I mean that without a good plan and a good architecture, the ease of use in Azure makes it quite easy to make mistakes that can bite us later!

I am personally a huge Azure fan because of the flexibility of the platform, but I have seen customers make the move to Azure without giving due thought to the specific needs that Azure needs to fulfill for the organization, as well as the security/compliance needs that the solution needs to live up to and/or implement as part of the package. If you are a company or organization that has built their technology stack on Microsoft, then moving to Azure makes a lot of sense, but keep in mind that it is a completely different environment, with different security controls and different technologies running the Azure infrastructure!

N7 is already running M365 and Dynamics 365, making integration between these platforms and Azure a breeze. Microsoft has been putting in a lot of development efforts and investments into security and compliance. The integration between the Microsoft security and Compliance tools makes the decision of using Microsoft as the vendor for security a no-brainer for many organizations around the world. The Microsoft security and compliance tooling are equal to any of the other vendors out there, but with Microsoft, the tooling is integrated, making the reporting and monitoring of security with Microsoft tools second to none, in my opinion.

What kinds of security tools are available to us in Azure? Since Azure is a virtualized platform we can use 3rd party security tooling in Azure, but for the purposes of this book I will limit the list to the ones provided by Microsoft:

1. Azure Firewall
2. Web Application Firewall
3. Azure Security center
4. Azure Defender
5. Azure DDOS protection
6. Azure Security Posture Management
7. Azure Key Vault
8. Azure Active Directory...

Add to that, the many 3rd party integrations that can be used within an Azure subscription and you can see why a decision to move to Azure as a cloud provider should not be made on a whim. The above tools are used across Azure as a whole, but there is tooling that is applicable only to some of the Azure areas, like Azure SQL.

Azure SQL, as a data platform, comes with most of the security features of the on-premises SQL Server database, but in addition to those Azure SQL comes with a feature that is unique to the Azure cloud:

1. TLS Encryption for data in transit
2. Advanced Threat Protection – Unusual behavior monitoring
3. Data Discovery and Classification

The same goes for many of the other offerings we can use on the Azure cloud, like:

1. Azure Storage
2. Azure Batch
3. Azure Kubernetes Services
4. Azure Functions
5. Azure Logic Apps

6. Azure Web
7. Azure Table Storage...

The above list has not even begun to list all the offerings that Microsoft has on Azure. Add to those the many 3rd party integrations and you can see why many companies chose Microsoft Azure as the cloud platform of choice, but hopefully, you can also see why many organizations quickly get themselves painted into a corner, because of the complexity and amount of offerings on Azure.

10.8 Final Thoughts

We have now reached the concluding section of this book on Security Architecture. I hope that you have gained additional knowledge that you can apply in your day-to-day work situation. I am aware that the first nine chapters were highly theoretical in nature, unfortunately, they had to be before we could get to a more concrete example in this chapter.

Undoubtedly, some of you have just read chapter ten, with the N7 example on a security architecture development case, which is fine, make no mistake! I hope you will return to the theoretical chapters at a later point, as they give you a foundational knowledge that you can use to get the SABSA foundational certification for instance, but the theory will also give you the knowledge to adapt SABSA to real-world situations you will have to apply SABSA in.

One of my favorite quotes, related to architecture, is a quote I read on the Internet. It goes like this:

'If you think good architecture is expensive, you should try bad architecture'

That exact same quote can be applied to security architecture. Designing and implementing a security architecture will be a task that requires both time and resources to implement in a fashion that will be effective for the organization. But and there is always a but, the benefits gained from having a well-designed and maintained security architecture will save money while providing the organization with enough agility to adapt to new situations while maintaining a security posture that can withstand the complex and ever changing threats that are present on the Internet and within IT in general.

Index

About the Author

Tom Madsen works as a Cybersecurity SME for NNIT in Denmark, specializing in security & Compliance from Microsoft, Cisco & Oracle. Author of the book Sun Tzu – The Art of war for Cyber Security.

For Product Safety Concerns and Information please contact our EU
representative GPSR@taylorandfrancis.com
Taylor & Francis Verlag GmbH, Kaufingerstraße 24, 80331 München, Germany

www.ingramcontent.com/pod-product-compliance
Ingram Content Group UK Ltd.
Pitfield, Milton Keynes, MK11 3LW, UK
UKHW021110180425
457613UK00001B/12